THE PHYSICAL METALLURGY OF
TITANIUM ALLOYS

ASM Series in Metal Processing

Series Editor

Harold L. Gegel, FASM

Senior Scientist
Air Force Wright Aeronautical Laboratories

THE PHYSICAL METALLURGY OF
TITANIUM ALLOYS

E.W. Collings
Senior Research Scientist
Battelle Memorial Institute
Columbus, Ohio

American Society for Metals
Metals Park, OH 44073

First printing, September 1984
Second printing, October 1986

Library of Congress Cataloging in Publication Data
Collings, E. W.
   Physical metallurgy of titanium alloys.

   (ASM series in metal processing)
   Includes index.
   1. Titanium alloys.   I. Title.   II. Series.
TN693.T5C63 1984   669′.967322   84-2838
ISBN 0-87170-181-2

SAN 204-7586

PRINTED IN THE UNITED STATES OF AMERICA

To the members of R.I. Jaffee's
Metal Science Group at Battelle
among whom I first became acquainted
with titanium and its alloys

Technical Illustrations: Judith S. Ward

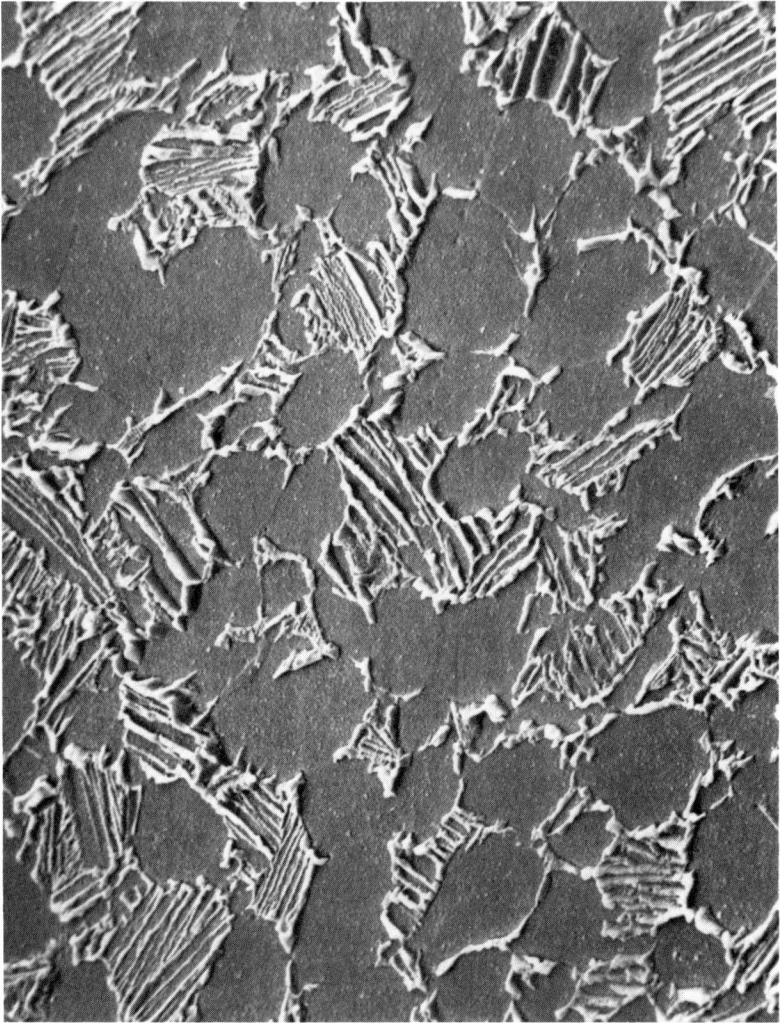

Alpha plus transformed-beta structure of the alloy Ti-6.1Al-2.1Sn-3.9Zr-2.1Mo-0.084Si (i.e. Ti-6242-0.1Si) in the condition: final-forged at 1750°F (955°C) to an area reduction ratio of 2.8:1 then annealed 2h/1775°F (970°C) and air cooled. Mounted in the longitudinal section. Specimen courtesy of S.L. Semiatin, optically polished and deep etched (5 vol.% HNO$_3$, 2$\frac{1}{2}$ vol.% HF, bal. aqu.) by C.R. Thompson, and photographed in the scanning electron microscope by A. Skidmore (all of Battelle). Present magnification, 1440×. The same sample is also shown in *Fig. 3-18(b)*.

Foreword

The American Society for Metals is responding to the technical needs of a growing fraternity of scientists, engineers, and production managers of the second- and third-tier industrial base who are being challenged by rapid advances in the technology of automation.

The introduction of computer-aided and robotic technologies into the manufacturing process has caused managers to rethink manufacturing operations in terms of providing the full market value of which they are capable. The up-to-date factory is striving to reach the "ideal" goal of stockless production, which can be achieved only through closer relationships between design and manufacturing and between the designer and the parts vendor. In general, a productivity gap has always existed between vendors and users, because vendors traditionally have not been partners with their customers. The parts and material suppliers must now improve the quality of their products and their productivity to bring their production capabilities in line with the new trends in world manufacturing.

The quality of products produced by the factory of the future will depend on the parts vendor's ability to change from a job-shop mode of production to one of zero lead time with repetitive production of discrete units that are free of defects.

The ASM Process Modeling Activity members are formulating a series of monographs that will link the components of process modeling and focus on its use to facilitate modernization of the vendor industries served by the society.

Although the range of process modeling is very broad, these monographs will focus primarily on the metalworking community. They will be aimed at supporting the continuing education of graduate engineers, and each monograph will have sufficient technical depth to be useful as reference material for university courses at the graduate and upper-undergraduate student levels. The monographs will link equipment or machinery dynamics, material-behavior modeling, numerical methods for process simulation, interface phenomena, and process economics. One of the aims of this series is to underscore the fact that process modeling is intimately associated with process simulation, die (mold) design and manufacturing, and process control.

The theory and examples contained in these monographs provide the basis for understanding the value of computer-aided engineering and manufacturing (CAE/CAM) as a productivity and part-quality enhancement tool. CAE/CAM systems will serve as a direct link between the designer and the parts vendor, making the vendor a "partner" in the total manufacturing operation.

In preparing these monographs, the authors have drawn extensively from the existing literature and the published results of active research programs in processing science. In particular, the Air Force Office of Scientific Research (AFOSR) program on processing science has had significant influence. This research encompassed materials, mechanics, interface phenomena, and equipment characteristics in a unified approach to the modeling of deformation processes. The various components of the modeling system were integrated by means of an interactive computer program that facilitates the simulation of material flow in arbitrarily shaped dies. The analytical models make possible the design of dies based on metal-flow simulation and the development of a process-control algorithm for feedforward control of metalworking presses; furthermore, they provide the means for using process control as a method of guaranteeing quality assurance of net-shape products. Shop-floor validation of process modeling for the design of near-net-shape products having a controlled set of microstructures and properties has been a rewarding result of this research.

It is a pleasure to acknowledge the support provided by AFOSR and, in particular, Dr. T.E. Walsh, Electronic and Materials Science, for his continued support. The stimulation and guidance provided by Dr. Frank Kelley, Dr. H.M. Burte, Dr. N.M. Tallan, and colleagues of the Air Force Wright Aeronautical Laboratories/Materials Laboratory throughout the program is also gratefully acknowledged. Special thanks are also due to Dr. George Dieter, Dean of the College of Engineering at the University of Maryland, and Dr. Roger N. Wright, Professor at Rensselaer Polytechnic Institute, who provided consultative advice. Special recognition must be given to Dr. Shiro Kobayashi, Professor of Mechanical Engineering at the University of California — Berkeley, for his dedicated research which has significantly advanced the analysis of metalworking processes. Finally, it is a pleasure to acknowledge Dr. Taylan Altan, of Battelle Columbus Laboratories, who was the principal investigator for the AFOSR processing science program.

Harold L. Gegel
Series Editor

Preface

Origin and Uses of Titanium

Titanium is widely distributed throughout the universe. It has been discovered in the stars, interstellar dust, meteorites, and on the surface of the earth. Its concentration within the earth's crust of about 0.6% makes it the fourth most abundant of the structural metals (after Al, Fe and Mg). It occurs twenty times more frequently than Cr, thirty times more than Ni, sixty times more than Cu, one hundred times more than W, and six hundred times more than Mo. This abundance is to some extent illusory, however, in that titanium is not so frequently found in economically extractable concentrations. Concentrated sources of the metal are the minerals ilmenite, titanomagnetite, rutile, anatase, and brookite.

Ilmenite is haematite (Fe_2O_3) in which half of the Fe has been replaced by Ti; titanomagnetite is magnetite (Fe_3O_4) in which one-third of the Fe has been replaced by Ti. Rutile is TiO_2 (as are anatase and brookite). Naturally occurring (and Ti-deficient) ilmenite consists of haematite particles in a matrix of ilmenite; naturally occurring (and, again, Ti-deficient) titanomagnetite is magnetite containing laths of ilmenite. In short, the most important titanium minerals are ilmenite and rutile.

Titanium was first discovered in minerals now known as rutile by W. Gregor (England) and M.H. Klaproth (Germany) in about 1790. The first commercial mill products were produced by the Titanium Metals Company of America (TMCA) around about 1950. From that time to the present, production of the metal has grown at an average rate of about 8% p.a. Superimposed upon part of this temporal growth curve is a large fluctuating component, a reminder of the capriciousness of the materials demands of the aerospace industry, titanium's principal market during the early years. Fortunately for the titanium-production industry, the 13% p.a. growth rate exhibited by the civilian sector of the total market since the early 1960s has served to more than offset the decline in military demand during the same period, thereby yielding not only a net growth but a relatively steady one.

In order to cope with unexpected increases in the demand for a metal it is helpful to be able to rely on a copious and stable supply of the basic ore. The titanium industry is fortunate in this regard. Titanium dioxide is produced in large quantities for many applications, so much so that in 1977, for example, only a few percent of the world's production of titanium ore was tapped for metallic sponge refinement. Thus, since the overall demand for raw material is not subject to the same fluctuations as the demand for the metal, should the latter undergo a significant increase at any time there is at least a strong raw-material base from which to draw.

Industry's growing awareness of the need for energy conservation has served to emphasize an unfortunate characteristic of the current methods of titanium-metal refinement: their energy intensiveness. The energy required to produce a ton of sponge-Ti from its ore is 16 times that needed to produce a ton of steel, 3.7 times that needed to produce the same amount of refined Cu, 3.2 times that needed for ferrochrome, 1.7 times that needed for Al production, and a little more than that needed for a 1-ton ingot of Mg. Since, however, the heats of formation of rutile (~ -228 kcal mole^{-1}), haematite (~ -200 kcal mole^{-1}), and magnetite (~ -268 kcal mole^{-1}) are in the ratio of $1/0.88/1.18$, there seems to be some scope for increasing the energy efficiency of the titanium-refinement process.

Titanium (meaning Ti and its alloys) has two principal virtues: (i) a high strength/weight ratio, and (ii) good corrosion resistance. At one time or another practically all aerospace structures — air frames, skin, engine components — have benefited through the introduction of titanium. Nonaerospace applications include: steam-turbine blades, hydrogen-storage media, high-current/high-field superconductors, condenser tubing for nuclear and fossil-fuel power generation, and other corrosion-resistant applications such as components for ocean thermal-energy conversion, off-shore oil drilling, marine-submersible vessels, desalinization plants, waste-treatment plants, the pulp-and-paper industry, and the chemical and petrochemical industries.

Important Literature Sources

The emerging importance of the metal as described above has given rise to, and in turn been aided by, the appearance of numerous monographs and conference proceedings devoted entirely to descriptions of the properties of titanium and its alloys. A *Handbook on Ti* authored by H.K. Adenstedt appeared in 1954 (Part I) and 1955 (Part II) (see the accompanying reference list in this section). These were followed by an impressive and well-written comprehensive treatise on *Titanium* published in

1956 under the authorship of A.D. and M.K. McQuillan. Important reviews of "The Physical Metallurgy of Titanium Alloys" and "Titanium Metallurgy" were presented, respectively, by R.I. Jaffee in 1958 and H. Margolin and J.P. Nielsen in 1960. Soviet contributions to the science and technology of titanium were described in a series of ten volumes entitled *Titanium and Its Alloys* [the second (1962) and the tenth (1963) being the proceedings of conferences] edited by I.I. Kornilov. The physical metallurgy and metallurgy of titanium as of ca. 1973 have been fully described in a monumental work by U. Zwicker entitled *Titan und Titanlegierungen*. The nonaerospace applications of titanium, representing an extensive and growing civilian market, have been discussed in a book published in 1981, edited by D. Eylon, entitled *Titanium for Energy and Industrial Applications*. Finally, attention is drawn to a major source of information on the scientific and technological properties of titanium — the proceedings of the four international conferences which have so far been held on the subject and which were published in 1970 (the first conference, London, one volume), in 1973 (the second conference, Cambridge, MA, four volumes), in 1980 (the fourth conference, Kyoto, Japan, four volumes), and in 1982 (the third conference, Moscow, three volumes).

The Science and Technology of Titanium Alloys: Literature Sources

- ADENSTEDT, H.K., *Handbook on Ti*, Wright Air Development Center Tech. Report No. 54-305; Part I, Aug. 1954; Part II, Sept. 1955.

- McQUILLAN, A.D. and McQUILLAN, M.K., *Titanium*, Academic Press and Butterworths Scientific Publications, 1956.

- JAFFEE, R.I., The Physical Metallurgy of Titanium Alloys, *Progr. Metal Phys.*, 7, 65-163 (1958).

- MARGOLIN, H. and NIELSEN, J.P., Titanium Metallurgy, in *Modern Materials, Advances in Development and Application*, Vol. 2, ed. by H.H. Hausner, Academic Press, 1960, pp. 225-325.

- KORNILOV, I.I. (ed.), *Titanium and Its Alloys*, in particular, publication No. 10, transl. from the Russian by Israel Program for Scientific Translations, Jerusalem, 1966.

- ZWICKER, U., *Titan und Titanlegierungen*, Springer-Verlag, 1974.

- EYLON, D. (ed.), *Titanium for Energy and Industrial Applications*, The Metallurgical Society of AIME, 1981.

- JAFFEE, R.I. and PROMISEL, N.E. (eds.), *The Science, Technology, and Application of Titanium* (Proc. 1st Int. Conf. on Titanium, London), Pergamon Press, 1970.

- JAFFEE, R.I. and BURTE, H.M. (eds.), *Titanium Science and Technology* (Proc. 2nd Int. Conf. on Titanium, Cambridge, MA), Plenum Press, 1973.

- WILLIAMS, J.C. and BELOV, A.F. (eds.), *Titanium and Titanium Alloys, Scientific and Technological Aspects* (Proc. 3rd Int. Conf. on Titanium, Moscow), Plenum Press, 1982.
- KIMURA, H. and IZUMI, O. (eds.), *Titanium '80: Science and Technology* (Proc. 4th Int. Conf. on Titanium, Kyoto, Japan), The Metallurgical Society of AIME, 1980.

Genesis and Scope of This Book

It is hoped that the present volume will be judged to be a worthy companion to its predecessors. It has been written from the point of view of an author who tries to view the natural world from as fundamental a scientific standpoint as possible. This book has been dedicated to the members of the Metal Science Group at Battelle, among whom the author first became interested in the properties of the binary research alloys of titanium when, in 1966, he was confronted with the task of describing in physical terms their mechanisms of solution strengthening and structural-phase stability. The model binary alloys selected for study at the time were Ti-Al (for solution strengthening) and Ti-Mo (for phase stability), systems which this book will go on to demonstrate are useful and valid prototypes of the technical α and metastable-β alloys, respectively. In agreement with the work of earlier authors, phase stability was shown to bear a one-to-one relationship to the β alloy's group number or electron/atom ratio. Solution strengthening, also treated from an electronic standpoint, was discussed within the framework of what might be termed a "short-range-order model" for dislocation pinning.

The electronic properties of those alloys were probed using physical-property measurements as diagnostic tools, at high as well as very low temperatures. The β alloys were, of course, superconductive in the liquid-He temperature range. Thus the measurement of the low-temperature specific heats of those alloys (in various thermomechanically processed conditions, as was appropriate to the task at hand) led naturally into a study of alloy superconductivity in general and eventually to the production of a book entitled *A Sourcebook of Titanium Alloy Superconductivity* (Plenum Press, 1983). The fundamental solution-strengthening work was carried out in close collaboration with H.L. Gegel of the US Air Force Materials Laboratory, Wright-Patterson Air Force Base, Ohio. It gave rise in due course to a TMS-AIME/ASM-sponsored symposium on the subject, the proceedings of which were published in a book entitled *Physics of Solid Solution Strengthening* (Plenum Press, 1975). It was most appropriate, therefore, for a book on titanium-alloy physical metallurgy to be commissioned for inclusion in this Metal Processing Monograph Series by Dr. H.L. Gegel, the Series Editor.

The purpose of this book is to provide a fundamental picture of the properties of binary research alloys in terms of which those of the technical alloys which derive from them can be easily understood. For example: (i) Ti-Al provides a model for solution strengthening; moreover, it is the prototype of technical ternary α alloys such as Ti-5Al-2.5Sn and the α components of $\alpha + \beta$ alloys. (ii) Ti-Mo provides models for both metastable and stable β alloys; a discussion of its properties aids in the understanding of those of the β components of $\alpha + \beta$ alloys as well as those of the technical metastable-β alloys themselves — Ti-11.5Mo-6Zr-4.5Sn (so-called "β-III"), for example.

An overview of the various classes of alloys to be discussed in this book is given in Chapter 1. Physical properties are considered in Chapter 2, which places emphasis on the roles that they play in studies of the statics and dynamics of microstructural change (such as precipitation and texturization). Chapter 3 deals with equilibrium phases. Again, some representative binary alloys set the stage for discussions of equilibrium phases in the technical alloys. Nonequilibrium phases are considered in Chapter 4; although both α and β alloys, when sufficiently dilute, can undergo martensitic transformation, only the β alloys exhibit ω phase and β' precipitation, and $\beta' + \beta''$ phase-separation, all of which have influenced, at one time or another, the properties of technical alloys. Chapter 5, the lengthiest chapter, deals with elastic and plastic mechanical properties. Included for discussion are the static (e.g., normal-tensile) and dynamic (e.g., ultrasonic) elastic moduli and the "normal" and "anomalous" classes of plastic properties. Bauschinger effect, for example, is treated in the latter category. Hardness, regarded as a "normal" plastic property, is extensively discussed in relationship to modulus and yield strength. The book concludes with a pair of chapters on deformation and aging, respectively. Featured in the first of these are discussions of superplasticity, transformation-assisted plasticity, and texturization; featured in the second are discussions of the effects of aging on the various fundamental microstructural phenomena described in Chapter 4 [e.g., martensite (α^m), ω phase, and β' precipitation] and, based on this, discussions of the effects of aging on the properties of several technical alloys.

Acknowledgements

The writing of this book was commissioned by Dr. H.L. Gegel and principally funded by the USAF Wright Aeronautical Laboratories, Wright-Patterson Air Force Base, Ohio, under Contract No. F33615-78-C-5025

("Research to Develop Process Models for Producing a Dual-Property Titanium Alloy Compressor Disc"; Dr. H.L. Gegel, AF Program Monitor) through the Battelle program managed by Dr. T. Altan. The support of both these scientists throughout the project is warmly appreciated. The original components of this review are the result of work supported over the years by the US Air Force Office of Scientific Research (Dr. A.H. Rosenstein, Program Manager) and by programs of the Air Force Materials Laboratory managed, again, by Dr. H.L. Gegel. Original data presented are the results of research conducted at Battelle with the assistance of Dr. J.C. Ho and Mr. R.D. Smith. The interpretation of the results, particularly those related to the solution strengthening of α alloys, was aided by discussions with Drs. H.L. Gegel and J.E. Enderby. The fundamental β-alloy work, particularly that relating to superconductivity, was performed in close collaboration with Dr. J.C. Ho. In the early stages of its preparation this manuscript benefited from the assistance of Drs. H.L. Gegel and S.L. Semiatin, both of whom supplied some useful source material, and of Dr. J.C. Williams, who provided numerous excellent photographs of titanium-alloy microstructures. Mr. R.A. Wood provided information on some of the commercial alloys of titanium. The manuscript was typed in final form by Mrs. J. Bulford. The technical illustration was performed by Ms. J.S. Ward.

E.W. Collings
Materials Department
Battelle Memorial Institute
Columbus, OH 43201
October, 1983

Contents

③ Equilibrium Phases   39

5 Mechanical Properties   111

7 Aging   181

Introduction

1.1 Titanium Alloys and Their Applications

Interest in the properties of Ti and its alloys began to accelerate during the late 1940s [CRA49] and early 1950s as their potential as high-temperature, high-strength/weight materials with aeronautical applications became more and more widely recognized. The history of Ti and its development in alloyed form has been described in detail in the introduction to the first International Conference on the subject [JAF70] and in the introduction to ZWICKER's comprehensive metallurgical treatise *Titan und Titanlegierungen* [ZWI74]. As evidenced by the papers presented at the subsequent International Conferences on Ti held in 1973 (Boston) [JAF73], in 1976 (Moscow) [WIL82], and in 1980 (Kyoto) [KIM80], Ti and its alloys have by now found widespread use in the aerospace industry (for both frame and engine components) and in the chemical and related industries where advantage can be taken of its corrosion resistance. According to WOOD [WOO72], by 1972 about thirty commercial alloys were already on the market in mill-product form. Of these, the eight most favored compositions, accounting for some 90% of the sales, were three grades of unalloyed Ti and the alloys Ti-5Al-2.5Sn, Ti-6Al-4V, Ti-8Al-1Mo-1V, Ti-6Al-6V-2Sn, and Ti-13V-11Cr-3Al. At that time also, interest in each of the alloys Ti-6Al-2Sn-4Zr-2Mo (i.e., "Ti-6242"), Ti-6Al-2Sn-4Zr-6Mo (i.e., "Ti-6246"), and Ti-11.5Mo-6Zr-4.5Sn (i.e., "β-III") was on the increase. Today the alloy Ti-6242 to which about 0.1% Si has been added is being used in Ti-alloy forgings and has received extensive study in its role as a candidate gas-turbine compressor-disc material. Finally it should be noted that Ti-10V-2Fe-3Al has been the beneficiary of the renewed interest currently being shown in so-called "near-β" Ti alloys [DUE80ª][TER80] [TOR80], while it is at last becoming recognized that Ti-50Nb, one of the most important of today's technical superconductors, is in fact a β-Ti alloy [COL81].

1.2 Classification of Titanium Alloys

Pure Ti undergoes an allotropic transformation from hcp (α) to bcc (β) as its temperature is raised through 882.5°C [Mol65][Zwi74]. Elements which when dissolved in Ti produce little change in the transformation temperature (e.g., Sn) or cause it to increase (e.g., Al, O) are known as "α stabilizers"; they are simple metals (SM) or the interstitial elements [Mol65, p. 154] – generally non-transition elements. Alloying additions which decrease the phase-transformation temperature are referred to as "β stabilizers"; they are generally the transition metals (TM) and noble metals – i.e., metals which, like Ti, have unfilled or just-filled d-electron bands. In the alloys, of course, the single-phase-α and single-phase-β regions are not in contact as they are in pure Ti; they are instead separated by a two-phase $\alpha + \beta$ region whose width increases with increasing solute concentration. Based on these considerations, technical alloys of Ti are classified as "α", "$\alpha + \beta$", and "β", *Fig. 1-1*.

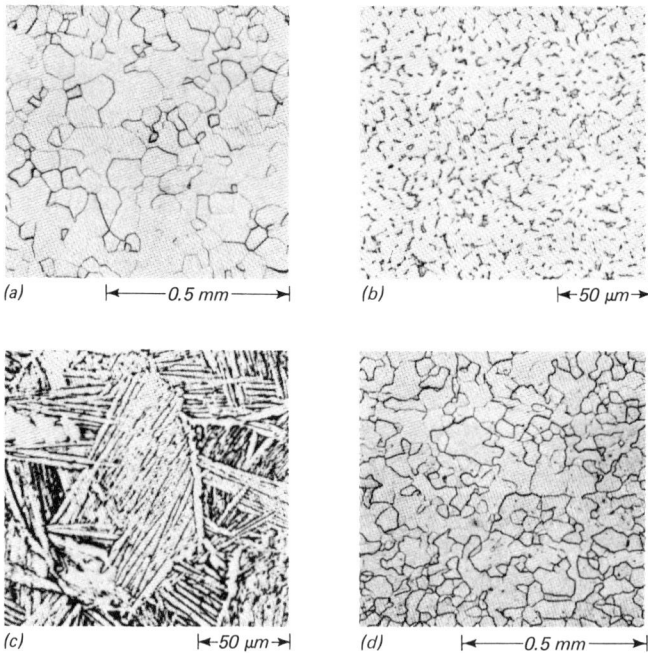

(a) |←——0.5 mm——→| (b) |←50 μm→|

(c) |←50 μm→| (d) |←——0.5 mm——→|

Fig. 1-1. Typical microstructures of α, $\alpha + \beta$, and β-Ti alloys: *(a)* equiaxed α in unalloyed Ti (after 1h/1290°F); *(b)* equiaxed $\alpha + \beta$; *(c)* acicular $\alpha + \beta$ in Ti-6Al-4V; *(d)* equiaxed β in Ti-13V-11Cr-3Al [Woo72, p. 1-0:72-4].

1.2.1 α Alloys

Unalloyed Ti and alloys of it with α stabilizers such as Al, Ga, and Sn, either singly or in combination as in the commercial alloy Ti-5Al-2.5Sn or the experimental Ti-Al-Ga alloys [Hoc73][Geg73], are hcp at ordinary temperatures and as such are classified as α alloys. These alloys, according to Wood [Woo72], are characterized by satisfactory strength, toughness, creep resistance, and weldability. Furthermore, the absence of a ductile-brittle transformation, a property of the bcc structure, renders α alloys (typified by Ti-5Al-2.5Sn) suitable for cryogenic applications [Sal79].

1.2.2 β Alloys

As mentioned above, transition-metal (TM) solutes are stabilizers of the bcc phase. Thus all-β alloys generally contain large amounts of one or more of the so-called "β-isomorphous"-forming additions (see Sect. 3.2) − V, Nb, Ta (group-V TMs), and Mo (a group-VI TM). The systematics of β stabilization in binary and multicomponent Ti-base alloys has been discussed in detail by Ageev and Petrova [Age70]. The archetypal binary β-stabilized Ti-base alloy, about which a great deal of physical and metallurgical information has been garnered over the years, is Ti-Mo. For a useful overview of the mechanical properties and aging characteristics of a pair of typical β alloys, Ti-15Mo-5Zr and Ti-15Mo-5Zr-3Al, the work of Nishimura *et al.* [Nis82] is recommended. There are several important commercial β alloys; one which has been attracting considerable attention recently is Ti-11.5Mo-6Zr-4.5Sn (β-III) [Fro73] [Pet73][Vig82][Wil82ᵃ]. β alloys, according to Wood [Woo72], are extremely formable. They are, however, prone to ductile-brittle transformation [Gor73] and, along with other bcc-phase alloys, are unsuitable for low-temperature applications [Sal79].

1.2.3 α + β Alloys

These are alloys whose compositions, usually at room temperature, are such that they support a mixture of α and β phases. Although many binary β-stabilized alloys in thermodynamic equilibrium are two-phase, in practice the α+β alloys usually contain mixtures of both α and β stabilizers. The simplest of such alloys, and one upon which the most attention has undoubtedly been lavished, is Ti-6Al-4V. Although this particular alloy is difficult to form, even in the annealed condition [Sal79], α+β alloys generally exhibit good fabricability as well as high room-temperature strength and moderate elevated-temperature strength. They may contain between 10 and 50% β phase at room temperature; if they contain more than 20%, they are not weldable. The properties of α+β alloys can be controlled by heat treatment, which is used to adjust the microstructural and precipitational states of the β component.

1.2.4 Summary

The mechanical properties and strengthening characteristics of Ti alloys have been reviewed within the framework of the above classification scheme by JAFFEE [JAF58] and subsequently by ZWICKER [ZWI74, pp. 248-305]. Some typical optical microstructures of α, $\alpha + \beta$, and β alloys were given in *Fig. 1-1*. A list of popular commercial alloys and their classifications according to that scheme is given in *Table 1-1*.

**Table 1-1. Structural Classes of
Commercial Titanium-Base Alloys
[Woo72][STR82]**

Alloy		Classification	
Ti-5Al-2.5Sn		α	
Ti-8Al-1Mo-1V	} near-α*		
Ti-6Al-2Sn-4Zr-2Mo			
Ti-6Al-4V			} $\alpha + \beta$
Ti-6Al-2Sn-6V			
Ti-3Al-2.5V			
Ti-6Al-2Sn-4Zr-6Mo	} near-β		
Ti-5Al-2Sn-2Zr-4Cr-4Mo			
Ti-3Al-10V-2Fe			
Ti-13V-11Cr-3Al			
Ti-15V-3Cr-3Al-3Sn			
Ti-4Mo-8V-6Cr-4Zr-3Al		} β	
Ti-8Mo-8V-2Fe-3Al**			
Ti-11.5Mo-6Zr-4.5Sn			

　　*The terms "lean-β" and "super-α" may also be used.
　　**Obsolete alloy.

1.3 Extraction of Titanium

The most well-known method of Ti production is the Kroll process, which involves the reduction of $TiCl_4$ by Mg. The first step in the process is the preparation of the tetrachloride itself, which is carried out by the chlorination of a mixture of carbon with the minerals rutile or ilmenite. The Kroll Mg-reduction reaction takes place in a closed heated reactor vessel under an inert atmosphere. Liquid $TiCl_4$ is introduced to the liquid Mg already present in the vessel, thereby initiating the reaction $2Mg + TiCl_4 \rightarrow 2MgCl_2 + Ti$. The reaction products are commercially pure sponge-Ti in the form of a grey porous coke-like mass and $MgCl_2$, most of which is able to be drained out of the reaction chamber as a liquid. The $MgCl_2$ is electrolytically recycled. The Ti sponge is consolidated by arc melting in a water-cooled Cu crucible—several iterations are performed of a procedure in which an arc is maintained between a consumable compacted-sponge-Ti electrode and a pool of molten sponge.

The highest purity Ti is prepared for research purposes by the iodide process. Crude Ti is first reacted with iodine in an inert atmosphere to form titanium iodide. This is then able to be decomposed at the surface of a heated Ti wire, which acts as nucleus for the growth of a long cylindrical bar of high-purity Ti crystals. Typical impurity contents of several grades of Ti are listed in *Table 1-2(a)* and *Table 1-2(b)*.

These and other standard commercial methods of Ti production such as the sodium-reduction (or Hunter) process, the direct-oxide-reduction process, and the electrolytic processes have been described in detail by the McQUILLANS [McQ56, Chap. 2], HOCH [Hoc73[b]], and ZWICKER [Zwi74, pp. 21-27], while some new approaches developed in the Soviet Union have been outlined by REZNICHENKO and co-workers [Rez82, Rez82[a]].

1.4 Practical Alloys for Low-Temperature and High-Temperature Applications

1.4.1 Low-Temperature Applications

As a consequence of their high strength/weight ratio, resistance to corrosion, and low thermal conductivities, Ti alloys are likely candidates for low-temperature structural applications. In practice, however, stainless steels are presently playing the dominant role in this regard. Single-phase-bcc alloys, because of their tendency toward low-temperature embrittlement, do not qualify for low-temperature service. On the other hand, suitable for a wide range of cryogenic applications are: unalloyed Ti (hcp), α-phase alloys such as Ti-5Al-2.5Sn, and $\alpha + \beta$ alloys such as Ti-8Al-1Mo-1V and Ti-6Al-4V [Hub73][Sal79]. The mechanisms of low-temperature strength and creep-resistance in α-Ti alloys have been considered in detail by CONRAD and co-workers [Con70] (see also [Con75] and [Con81]).

**Table 1-2(a). Total Impurity Contents
of Iodide- and Kroll-Process Titaniums
in Weight % [RAS72]**

Element	Iodide Ti	Kroll Ti
Mg	0.01	0.13
Si	0.01	0.05
Al	0.02	
Fe	0.01	0.20
Ni	0.01	
Co		0.02
Cr	0.01	
Mn	0.005	0.02
C	0.01	0.08
N	0.02	0.04
O	0.02	0.11

**Table 1-2(b). Typical Interstitial Impurity Contents of
Several Grades of Titanium**

Grade of titanium	Interstitial content, ppm			Data sources
	C	N	O	
MRC (MARZ-grade)	78	6	63	a
MRC (VP-grade)	150	40	350	b
TMC electrorefined sponge (grade ELXX)	---	40	370	c
Kroll-process (Toho sponge)	---	110	860	d
Kroll-process	800	400	1100	e
Iodide-process	100	200	200	e

(a) Materials Research Corporation: Zone-refined; supplied typical analysis.

(b) Materials Research Corporation: Vacuum-melted; supplied typical analysis.

(c) Titanium Metals Corporation: See also E.W. COLLINGS and J.C. Ho, *Phys. Rev.* **2**, 235-44 (1970).

(d) See E.W. COLLINGS and J.C. Ho, *Phys. Rev.* **2**, 235-44 (1970).

(e) See *Table 1-2(a)*.

1.4.2 Intermediate- and High-Temperature Applications

Whereas at low temperatures elements in substitutional solid solution are supposed to contribute to the athermal component of flow stress and interstitial elements to the thermal barriers, as the temperature rises *all* alloy species become more or less mobile and associate themselves with "atmosphere effects" to extents which depend on solute-atom diffusivities [Ros73]. Although chemical effects such as oxidation and hot-salt stress corrosion may limit the service temperature of a Ti alloy in some applications, the mechanical limitation is imposed by high-temperature creep. Thus the Larson-Miller creep parameter* becomes a useful index in terms of which the high-temperature dimensional stability of a Ti alloy may be characterized. Ti-6Al-4V, a popular general-purpose $\alpha + \beta$ alloy, was developed in the U.S. prior to 1955.** In subsequent years several alloys with improved mechanical heat resistance, in Larson-Miller terms, have been developed. Ti-6242, a forgeable alloy of considerable current importance, was introduced commercially in 1966. *Fig. 1-2*, from ROSENBERG's review of high-temperature Ti alloys [Ros73], traces the chronological development of heat-resistant Ti-base alloys. According to that article, the design of alloys for low- to intermediate-temperature service (less than 550°C) is based on the following general principles: athermal strengthening of the α phase is achieved by the addition of Al [SAR70] and the β

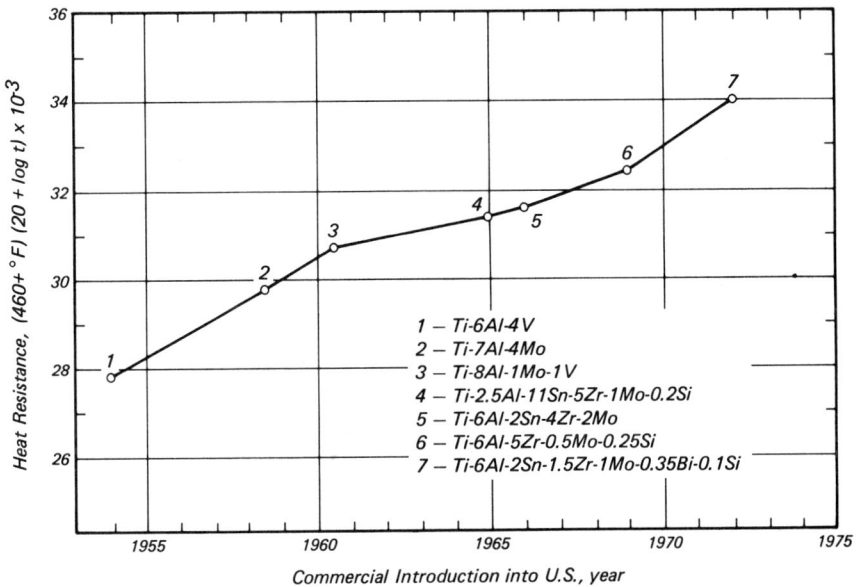

Fig. 1-2. Progress in the development of heat-resistant Ti-base alloys as gauged by the increase with time of the Larson-Miller parameter [Ros73].

---

*The Larson-Miller parameter represents the time to attain a specified strain (or to fracture) under a specified stress. It is given by $T(C + \log t)$, where T is the temperature in degrees Rankine (°F + 460), t is the time in hours, and C is a constant (about 20 for a large number of alloys) [HAY65, p. 135].

**Ti-6Al-4V accounted for 45% of the 1961 production of Ti, and 56% of the output in 1972 [Woo72].

phase by Mo [ZEY71]; the effect of Sn (and also Ga) is comparable to that of Al [COL75]; the effect of Zr was also at the time supposed to be comparable to that of Al (although the results of more recent work might take issue with that statement), with which it was supposed to interact to some extent.

By 1973, the Ti alloys most widely used in the U.S. at temperatures of 430°C and above were the "super-α" alloys: Ti-8Al-1Mo-1V and Ti-6242. Of these, the latter was the more heat resistant, being capable of service at temperatures of up to 480 ~ 510°C. Although several other compositions had been developed since 1966, when Ti-6242 was first introduced, they failed to achieve commercial acceptance either because of only marginal advantages over Ti-6242 or because, although exhibiting improved creep resistances, their stabilities were unsatisfactory [PAR73]. As indicated in *Fig. 1-2*, one alloy in particular, introduced commercially in 1973, did exhibit creep resistance markedly superior to that of its predecessor. Formed from Ti-6242 by the addition of Si to improve "surface stability"* and Bi to improve creep strength, this alloy, referred to as Ti-11, was intended for use in aircraft gas-turbine engines.

1.5 Classification of Heat-Resistant (High-Temperature) Titanium Alloys

Some principles which might be employed in the development of heat-resistant (i.e., high-temperature) Ti-base alloys and their classification into several microstructural categories have been described in detail by HOCH *et al.* [HOC73]. Attention was drawn to strengthening mechanisms operative in the following classes of alloy:

class 1: simple multicomponent α-phase solid solutions
class 2: simple $\alpha + \alpha_2$ two-phase systems
class 3: simple $\alpha + \alpha_2 + \beta$ + silicide systems
class 4: complex $\alpha + \alpha_2 + \beta$ + intermetallic-compound systems
class 5: α_2 systems
class 6: α_2 + intermetallic-compound systems
class 7: β systems (stable at all temperatures)
class 8: β + intermetallic-compound systems

Strengthening mechanisms supposed to be at work in these various alloy

---

*"Stability" refers to the maintenance of the original engineering properties, within prescribed limits, during exposure to the service environment at an elevated temperature under stress.

categories include: ordinary solid-solution strengthening (sss); sss augmented by short-range-order effects; sss augmented by hcp ordered-particle (α_2) effects; sss plus α_2 strengthening plus silicide or other intermetallic-compound precipitation within the β components of $\alpha+\beta$ alloys. Also considered by HOCH *et al.* as ingredients of high-temperature alloy design were "partitioning effects" associated with the addition of "supplemental elements" to a multicomponent, usually $\alpha+\beta$, base. For example, in the addition of Mo to, say, Ti-6Al-4V, since Mo is a stronger β stabilizer than V it will cause the latter to partition between the α and β components. The magnitude of this effect is highly sensitive to temperature, degree of deformation, cooling rate, etc. Important partitioning effects also take place in the presence of Si, which influences the extent to which oxygen divides itself between the α and β phases depending on the composition of the base alloy. The various interrelated effects which take place in the multicomponent alloy are too complicated to specify in detail in this space. The results of several studies, however, do indicate that multicomponent alloying cannot be treated from a simple additivity-rule standpoint but that atomic interactions may cause complicated changes within individual phases that could lead, on one hand, to the desired enhancement in strength [Hoc73, p. 11], but, on the other, to unexpected lattice instabilities [Hoc73, p. 12].

Physics and Physical Property Diagnostics

2.1 Introduction

2.1.1 Properties and Scope

Accompanying the development of Ti-base alloys and the associated mechanical-property measurements during the last 15 years have been a series of studies by COLLINGS and co-workers of one or more of the physical properties: electrical resistivity, Hall coefficient, magnetic susceptibility, and low-temperature specific heat. Representative α-phase alloys which have been subjected to these measurements were: Ti-Al, Ti-Ga, Ti-Sn, Ti-Al$_x$-Ga$_x$, Ti-Al-Ga (Ti$_3$X-type), and Ti-Al-Sn (Ti$_3$X-type). Representative β-phase alloys were: Ti-Mo, Ti-Mo-Al, Ti-Mo-Si, Ti-Mo-Fe, and Ti-Mo-Al-Fe. The numerical results themselves have been reported in [COL71a], while a critical interpretation of these and other results acquired during the 1970s have been presented in [COL80]. FISHER and co-workers have conducted an extensive series of studies of the elastic properties of monocrystalline Ti-base and other transition-metal alloy systems [FIS70, FIS70a, FIS75][KAT79, KAT79a]. The elastic moduli of polycrystalline alloys, in relation to their compositions and microstructures, have been studied by FEDOTOV and co-workers and the results published in a series of papers spanning the period 1963-1973 [FED63, FED64, FED66, FED73]. Elastic modulus may of course be regarded as either a physical or a mechanical property. Again, with considerable emphasis on microstructural states and phase transformations, various physical properties such as electrical resistivity and superconducting transition temperature have been studied by POLONIS, co-workers, and students on research alloys such as Ti-Nb, Ti-Cr, Ti-Mo, Ti-Mo-Al, Ti-Nb-Al, and others [CHA73, CHA74][LUH68, LUH69, LUH70, LUH70a, LUH71, LUH72][POL69, POL70, POL71]. With regard to the physical properties of *technical* Ti-base alloys, a compendium of properties such as electrical resistivity, specific heat, thermal conductivity, and thermal expansivity of unalloyed Ti, Ti-5Al-2.5Sn, Ti-6Al-4V, Ti-8Al-1Mo-1V, and Ti-13V-11Cr-3Al covering the temperature range from very low $(2 \sim 20\,\text{K})$ to room temperature has been assembled by SALMON [SAL 79].

Within the space of a single chapter it is obviously not possible to do full justice to the literature referred to above. Instead, a brief review based on a representative collection of papers is presented in which the emphasis is placed on the manner in which physical-property measurements may be used as indicators of: (i) microscopic and macroscopic metallurgical states of Ti alloys, and (ii) phase transformations and precipitation effects which take place in response to heat treatment. In other words, this chapter will take the form of a survey of physical-property measurements as they apply to metallurgical-property diagnosis.

Dynamic elastic modulus, regarded herein as a mechanical property, will be treated along with static elastic modulus and the plastic properties in Chapter 6. Properties included for discussion in this chapter are listed in the following five sub-sections.

2.1.2 Electrical Resistivity Measurement

Measurements of electrical resistivity as function of composition and temperature have provided useful metallurgical insights into certain strength and stability properties of α-phase and β-phase alloys. In α-Ti alloys (i.e., Ti-SM alloys where the solute is a so-called "simple metal"), a large specific solute resistivity (resistivity per at.% solute) is indicative of rapid solid-solution strengthening and is often accompanied by a rapid hardening coefficient. In β-Ti-TM alloys, an anomalous resistivity composition dependence is associated with the composition range over which isothermal and athermal ω phases are expected, with an anomalous resistivity temperature dependence *within* this composition range indicating the occurrence of reversible precipitation or associated structural fluctuations.

2.1.3 Magnetic Susceptibility Measurement

Magnetic susceptibility is the sum of numerous terms, one of which, χ_P, the Pauli paramagnetism, is proportional to the density-of-states at the Fermi level, $n(E_F)$, an important fundamental electronic property. But, ignoring its underlying significance, magnetic susceptibility has been used to (i) delineate phase boundaries in quenched Ti-SM alloys, (ii) investigate the $\alpha_2 \to \alpha$ (order-disorder) transformation in Ti-SM alloys, and transformations to the β phase in both Ti-SM and Ti-TM alloys, (iii) augment electrical resistivity in studying reversible ω-phase precipitation in the temperature range 150-300 K in quenched Ti-TM alloys, and (iv) monitor the course of ω-phase precipitation during the protracted moderate-temperature aging of initially quenched Ti-TM alloys.

2.1.4 Low-Temperature Specific Heat Measurement

The specific heat at low temperatures, C, is generally the sum of two components: γT and βT^3, where T is the absolute temperature, γ is the electronic specific heat coefficient (proportional to the density-of-states at the Fermi level, referred to above), and β, the lattice specific-heat coefficient, contains the Debye temperature, θ_D. In the case of Ti-TM alloys, a decrease of θ_D to low values, when plotted versus composition or electron/atom ratio, signifies lattice softening interpretable as a precursor to ω-phase precipitation. If the sample is a superconductor, another electronic property obtainable from low-temperature specific heat measurements is T_c, the superconducting transition temperature. Both γ and T_c, together with the total magnetic susceptibility, χ, have been used to monitor isothermal ω-phase precipitation during aging. The electronic component of the low-temperature specific heat of a pure unstrained single-phase superconductor undergoes a sharp discontinuous jump at the superconducting transition. If, as a result of inadequate quenching, deliberate aging, or mechanical deformation, the superconductive specific heat jump is severely rounded, the fitting of a distribution of sharp jumps to it can provide information relating to the microstructural constitution of a polyphase sample.

2.1.5 AC Impedance Measurement

Low-temperature calorimetry provides a contactless means of studying the superconducting transition temperature of a bulk sample. AC impedance measurement is another such technique. In this method, as applied by LUHMAN [LUH70] to the study of metallurgical effects in Ti-TM alloys, particularly Ti-Cr, the sample is surrounded by a coil connected to an oscillator adjusted to some convenient frequency, say 1 kHz. An electronic voltmeter placed across the coil gives an indication proportional to the impedance of the coil + sample. Since this is sensitive to the permeability (hence AC susceptibility, dM/dH) of the sample, the voltmeter reading responds to the transition from the superconducting to the normal state as the temperature of the sample is increased. LUHMAN and others (e.g., [LUH70]) have exploited this technique in a study of the microstructural responses of several Ti-TM alloys to variation of composition and thermal treatment. Their work on Ti-Cr, a β-eutectoid alloy, might be regarded as an indirect companion to the comparable series of calorimetric studies performed by COLLINGS and Ho on Ti-Mo, a related β-isomorphous alloy system.

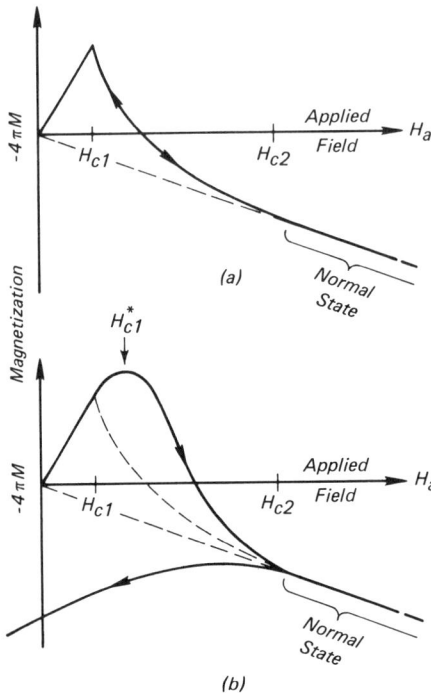

Fig. 2-1. Magnetization of a superconducting paramagnetic Ti-TM alloy — schematic diagrams of magnetization, $4\pi M$, versus the applied magnetic field, H_a: *(a)* reversible magnetization curve for an "ideal" or un-pinned annealed sample; *(b)* irreversible (hysteretic) curve for a sample with a high density of flux-pinning sites.

2.1.6 Magnetization Measurements of Superconducting β-Ti-TM Alloys

When a magnetic field, H_a, is applied to a type-II superconducting material, it is excluded from its interior by circulating surface super-currents until it reaches a value H_{c1}, the lower critical field. Penetration of the field to form what is known as the "mixed state"* then commences. As H_a increases, the normal fraction of the mixed state increases until the entire sample goes normal at H_{c2}, the upper critical field. If metallurgical defects of the kind which inhibit the ingress and egress of magnetic flux are absent, the magnetization is *reversible*, as in *Fig. 2-1*, curve *(a)*; on the other hand, if flux-pinning sites such as precipitates or other metallurgical irregularities are present, some of the applied flux will remain trapped when the applied field is removed — the magnetization is *irreversible*, *Fig. 2-1*, curve *(b)*. POLONIS and co-workers, particularly LUHMAN (e.g., [LUH70]), have employed this principle to study the precipitation of ω phase in Ti-Cr alloys as well as the $(\omega + \beta \rightarrow \beta' + \beta)$ ω-reversion effect.

---

*The mixed state is a microscopic ordered arrangement of normal and superconducting zones.

2.2 Electrical Resistivity

Studies of the composition- and temperature-dependences of electrical resistivity provide insights into strengthening mechanisms, phase stability, and the electronic structure of alloys. Employed as a diagnostic tool, electrical resistivity may be used to detect phase transformation during rapid quenching, and the measurement of relative resistivity during isothermal aging facilitates the construction of time-temperature-transformation (T-T-T) diagrams [SOE69][HOR73]. In studies of binary alloys of Ti, the favoring by solute atoms of either α-phase or β-phase stability subdivides Ti-base binary alloys into two classes: (i) alloys of Ti with simple metals or interstitial elements, and (ii) alloys of Ti with transition metals. *Fig. 2-2*, an example of this, shows the resistivity composition dependences of Ti-base alloys falling essentially onto two branches: an upper branch consisting of the Ti-SM and a lower branch corresponding to the Ti-TM alloys. As demonstrated by COLLINGS *et al.* [GEG73ᵃ][COL75ᵃ] and pointed out by STERN [STE75], the rapid strengthening exhibited by simple metals in Ti is consistent with their being strong scatterers of the conduction electrons.

The resistivity of an alloy can be usefully separated into two terms, thus:

$$\rho_{total} = \rho_i + \rho_s \ , \tag{2-1}$$

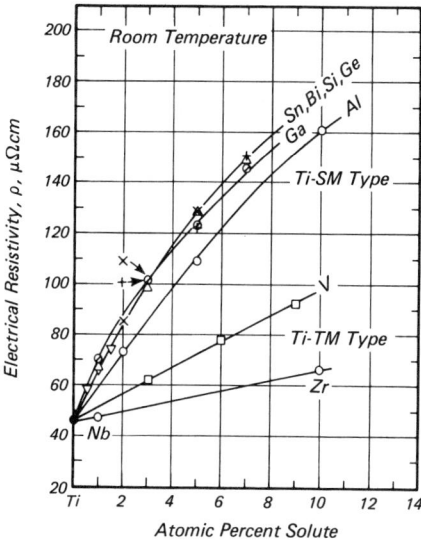

Fig. 2-2. Intercomparison between the composition dependences of electrical resistivity of two classes of Ti-base binary alloy: the Ti-SM type and the Ti-TM type. *Conditions*: as-cast: Ti-Sn ($\triangle$), Ti-Ga ($\oslash$), Ti-Al ($\circ$); 1h/1000°C/WQ: Ti-Ge ($\times$), Ti-Bi (+), Ti-Si ($\triangledown$), Ti-V ($\square$) [COL73ᵃ, COL75ᵃ].

where ρ_i, the "ideal" resistivity of the host, which may at high temperatures be expressed in the form [MEA64, p. 98]

$$\rho_i \rightarrow \frac{k_B T}{\theta_D^2} \, , \tag{2-2}$$

is a function of both the electronic structure of the alloy and thermal scattering, and the other term, ρ_s, represents the temperature-independent impurity scattering from the solute atoms.

In numerous low- or intermediate-concentration alloys it has been discovered that ρ_i and ρ_s are independent. Evidence in support of this property, known as Matthiessen's rule, is the parallelism frequently noted among the $\rho(c)$ curves for members of an alloy series. Naturally Matthiessen's rule breaks down when the presence of solute begins to influence ρ_i through its effect on $n(E_F)$ and θ_D, or for other reasons such as:

(a) When ρ_{total} becomes sufficiently large, as a result either of impurity scattering (ρ_s at high solute concentrations) or thermal scattering (ρ_i at high temperatures), further increments of solute or temperature, respectively, become relatively less effective. For example: (i) the specific resistivities of Al, Ga, Ge, or Sn in Ti-Mo(25 at.%) are on the average four times smaller than when the same elements are dissolved in pure Ti [GEG73ª][COL75ª]. (ii) At high temperatures the resistivity temperature dependences of some pure metals and alloys develop negative curvatures (see *Fig. 2-3*).

(b) The concentration dependence of resistivity of concentrated simple binary alloys not only decreases with increasing concentration, but passes through a maximum according to Nordheim's rule [MEA65, p. 113], which states:

$$\rho_s \propto c(1 - c) \tag{2-3}$$

(c) Numerous Ti-SM and Ti-TM alloys exhibit negative temperature coefficients of resistivity. Such gross departures from Matthiessen's rule require for their explanations detailed knowledge of the electronic structures and/or the phonon spectra of the alloys concerned.

2.2.1 Anomalous Resistivity Temperature Dependence, $d\rho/dT$

Negative resistivity temperature dependence has attracted considerable attention over a prolonged period of time. Depending on temperature range and alloy type, the phenomenon has been attributed to: (i) the Kondo effect (dilute alloys at low temperatures [RIZ74]), (ii) the increase with decreasing temperature of spin-disorder scattering from local moment

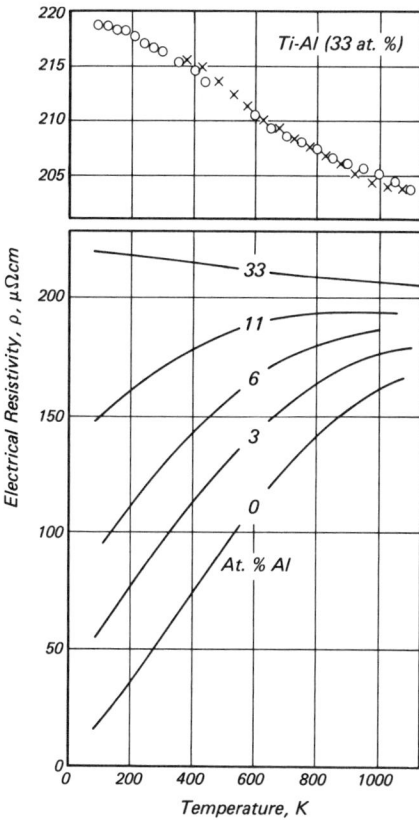

Fig. 2-3. Temperature dependences of the electrical resistivities, ρ, of unalloyed Ti and four Ti-Al alloys showing the tendency for $d\rho/dT$ to shift from strongly positive to weakly negative with increasing Al content [Moo73].

clusters (e.g., concentrated Cu-Ni alloys [Hou70]), (iii) an increase with decreasing temperature of the density of ω-phase precipitation itself [Ho72][Col74, Col78] (as in Ti-V and Ti-Mo alloys and related alloy systems — see also references in [Cha74]), and (iv) a smearing-out with increasing temperature of the density-of-states structure near E_F in certain classes of strong-scattering concentrated binary alloys [Che72]. Mechanism "iii", which in the spirit of the above three equations refers to the scattering contribution, ρ_s, and mechanism "iv", which relies on an alloy density-of-states effect, are of particular significance in this context, the former being applicable to Ti-TM alloys and the latter to Ti-SM.

2.2.2 Anomalous $d\rho/dT$ in Ti-SM Alloys

Ti-SM systems are strong-scattering alloys whose density-of-states functions possess considerable structure. Chen et al. [Che72], using the coherent potential approximation (CPA) [Fau82], have performed a model calculation on a concentrated binary alloy system, and have watched the changes in density-of-states, $n(E)$, which occur in response to: (i) change of solute concentration, (ii) change of solute scattering strength, or (iii) change of temperature. In order to do so they have calculated the relative electrical conductivity as a function of band filling, and have been able to predict in a semiquantitative way the manner in which resistivity may change with temperature in two classes of concentrated binary alloys: (i) *virtual-crystal or weak-scattering alloys*, characterized by featureless parabolic $n(E)$ curves, whose resistivities increase with temperature in the "usual way", and (ii) *strong-scattering alloys*, whose $n(E)$ curves possess deep minima or "pseudogaps", such that alloys whose compositions fall within the gap—which broadens and fills-in with increasing temperature—have electrical conductivities which increase with temperature (i.e., negative $d\rho/dT$'s). The salient features of the model are illustrated in *Fig. 2-4* and *Fig. 2-5*.

2.2.3 Anomalous Isothermal Resistivity Composition Dependence in Ti-TM Alloys

The resistivities of Ti-TM alloys exhibit isothermal resistivity-composition-dependence anomalies within which anomalous (i.e., negative) resistivity temperature dependences are located. The resistivity composition dependences of Ti-V, for example, at the temperatures 300, 200, and 77 K [COL74] are shown in *Fig. 2-6*. The corresponding quenched microstructures are also indicated in that figure. According to McCABE and SASS [MCC71], who have made a detailed TEM study of the system, ω phase is seen as a submicroscopic precipitate in the concentration range 13 through 25 at.% V, just that which includes the resistivity maximum. But although the sequence of sharp, then diffuse, electron-diffraction spots is confined

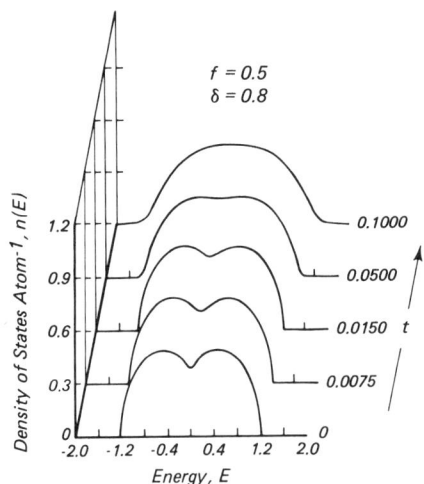

Fig. 2-4. Density-of-states, $n(E)$, versus energy, E, curves calculated using the CPA method for a model equi-atomic ($f = 0.5$) strong-scattering (parameterized by δ, with $\delta = 0.8$ on a scale of 0 to 1) binary alloy. Results for five values (0 to 0.1) of a reduced temperature, t, are indicated [CHE72].

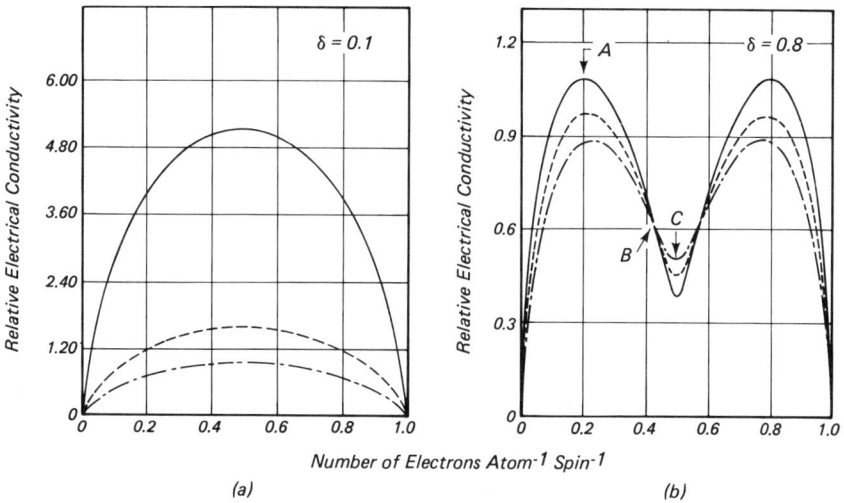

Fig. 2-5. Relative electrical conductivity as a function of band filling at three reduced temperatures for the model equi-atomic alloy of the previous figure: *(a)* weak-scattering case ($\delta = 0.1$); $t = 0.000$ (——), 0.006 (- - -), 0.012 (— - — - —); *(b)* strong-scattering case ($\delta = 0.8$); $t = 0.000$ (——), 0.0075 (- - -), 0.015 (— - — - —). In the strong-scattering case, which applies to *Fig. 2-3*, three $d\rho/dT$ signatures are possible depending on the level of band filling: $d\rho/dT$ is positive at A, zero at B, and negative at C [CHE72].

to the above concentration range, diffuse haloes persist in gradually decreasing intensity all the way across to pure V, a manifestation of a corresponding gradually decreasing lattice instability. The obvious conclusion is that the anomalous excess isothermal resistivity is closely associated in some way with the presence of both the athermal *and* the diffuse ω phases.

2.2.4 Anomalous $d\rho/dT$ in Ti-M Alloys

Fig. 2-6 shows, in addition to the effect considered above, that the resistivity isothermals intersect in such a way as to establish a negative $d\rho/dT$ within the composition interval 20 to about 33 at.% V. Presented in this way it appears that negative $d\rho/dT$ in alloys such as Ti-V is a minor perturbation of a much larger effect — the anomalous composition dependence — and, as such, is also related to the ω instability. Other Ti-TM systems in which negative $d\rho/dt$ has been studied are Ti-Nb [AME54] [PRE74], Ti-Cr [LUH68][CHA73, CHA74], Ti-Mo [YOS56][HAK61][HO72] [CHA73, CHA74], and Ti-Fe [HAK61][PRE76]. The question inevitably arose as to whether the negative $d\rho/dT$ was a consequence of reversible (athermal) ω-phase precipitation (as suggested in [HO72]) or a manifestation of the soft-phonon instability that gives rise to it [COL74]. Circumstantial evidence which could be taken in support of the former hypothesis

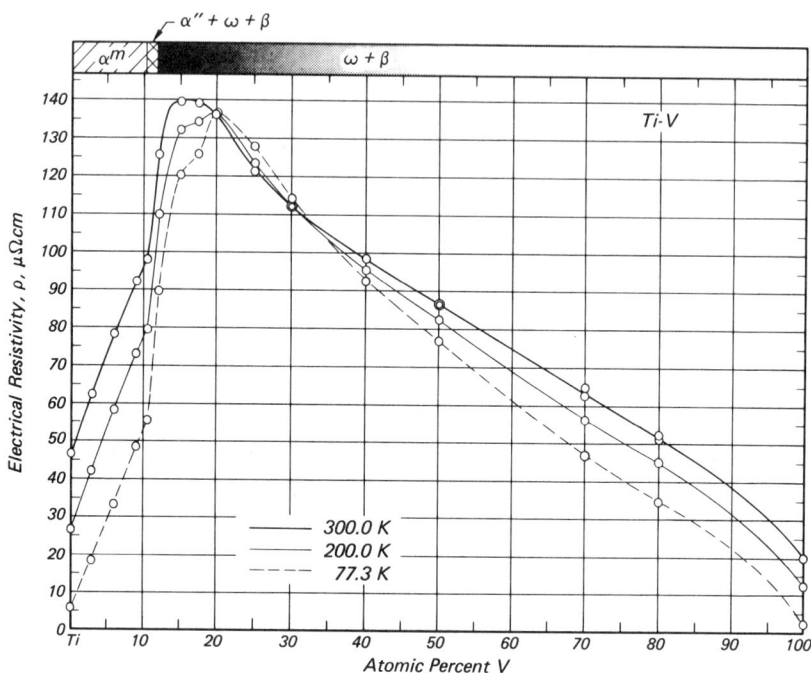

Fig. 2-6. Electrical resistivities of Ti-V alloys at three temperatures. Resistivities were measured at 77.3 K, 200 ± 1 K, and 298 ± 1 K. In the latter cases they were corrected to 200.0 K and 300.0 K, respectively, using measured $d\rho/dT$ data. Negative $d\rho/dT$ is found within the composition range 20 ~ 30 at.% V between the points of intersection of the isothermals [COL74].

can be presented in the form of *Fig. 2-7*, in which the anomalous reversible resistivity component, $\Delta\rho|^{300}_{77K}$, is juxtaposed against $\Delta f_\omega|^{300}_{150K}$, a magnetically deduced reversible change of crystalline athermal ω-phase abundance. But since the athermal ω is expected to be associated with a fluctuation (or diffuse) component, the result was still inconclusive. The picture has finally been clarified by POLONIS *et al.* [CHA74] in an elegant series of experiments commencing with measurements on quenched Ti-Cr(20 at.%). Since both the as-quenched $\omega+\beta$-phase alloy and the 435°C-reverted $\beta'+\beta$-phase alloy shared the same negative value of $d\rho/dT|^{273}_{77K}$, it became abundantly evident that the negative resistivity temperature dependence exhibited by Ti-Cr and similar alloys was associated with the ω-phase *instability* of the β phase itself, rather than its by-product.

2.3 Magnetic Susceptibility

As with many other physical properties, magnetic susceptibility not only provides useful information relating to the electronic structures of metals

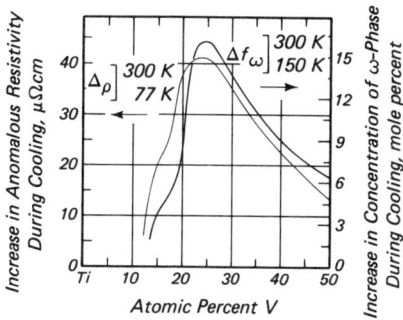

Fig. 2-7. Increase in anomalous resistivity, $\Delta\rho$, incurred on lowering the temperature of quenched Ti-V alloys from 300 to 77 K, compared with a magnetically derived estimate of the increase in ω-phase abundance which takes place as the temperature is lowered from 300 to 150 K [Col74, Col78].

and alloys but, when calibrated against suitable metallographic bench marks, can aid in phase-diagram investigation and in the interpretation of aging experiments. As an ingredient of the theory of superconductivity, the magnitude of the Pauli spin susceptibility component of the total normal-state susceptibility may profoundly influence the value of the mixed-state upper critical field.

2.3.1 The Total Magnetic Susceptibility

(a) Components of the Total Susceptibility. It is convenient to treat the total magnetic susceptibility of a transition metal or alloy as the linear superposition of components representative of contributions to it from (i) electrons at the Fermi level, (ii) states within the band, and (iii) the individual ion cores thus:

$$\chi = \underbrace{\chi_P + \chi_L}_{\text{(i)}} + \underbrace{\chi_{so} + \chi_{orb}}_{\text{(ii)}} + \underbrace{\chi_i}_{\text{(iii)}} ,$$

(2-4)

where the terms are entitled, respectively, Pauli spin paramagnetism, Landau diamagnetism, spin-orbit susceptibility, orbital paramagnetism, and ion-core diamagnetism. The properties of these individual components have been adequately discussed elsewhere [Col70, Col71, Col80]. For the present purpose it is sufficient simply to regard the total susceptibility as a macroscopic physical measurable and to consider its changes in response to changes of metallurgical variables.

(b) Magnetic Diagnostic Methods. The total magnetic susceptibility of a system of two components A and B of susceptibilities, $\chi_{A,B}$, and relative abundances, $f_{A,B}$ (with $f_A = 1 - f_B$), is given by the usual continuity equation:

$$\chi = f_A \chi_A + f_B \chi_B .$$

(2-5)

If χ_A and χ_B are known as a result of some preliminary investigation, Eqn. (2-5) can be manipulated so as to yield quantitative information relating to various metallurgical effects, processes, and properties such as (i) athermal ω-phase precipitation [COL78], (ii) the precipitation of ω phase during isothermal aging [COL75^b], and (iii) the development of equilibrium phase diagrams [COL79]. Finally, and in a somewhat different vein, advantage may be taken of the magnetic anisotropy characteristic of α-phase Ti-base alloys in order to quantify the crystallographic textures which they acquire as a result of anisotropic cold deformation. Alloys which have been examined in this way are Ti-Al(0, 3, 5, and 10 at.%) [COL82].

2.3.2 Magnetic Studies of Athermal ω-Phase Precipitation

An expression for the temperature dependence of the total magnetic susceptibility can be obtained by differentiating Eqn. (2-5). Performing this differentiation, and writing ω for A and β for B, we find:

$$\frac{d\chi}{dT} = \underbrace{f_\omega\left(\frac{\partial\chi}{\partial T}\right)_\omega + f_\beta\left(\frac{\partial\chi}{\partial T}\right)_\beta}_{(a)} + \underbrace{(\chi_\omega - \chi_\beta)\left(\frac{\partial f}{\partial T}\right)_\omega}_{(b)} \qquad (2\text{-}6)$$

The first pair of terms, (a), on the RHS of the equation are equivalent to $\partial\chi/\partial T$, the intrinsic temperature dependence of the total mean susceptibility. The second pair, (b), represents the change in susceptibility which takes place during reversible $\omega \rightleftarrows \beta$ allotropic transformation. f_ω, the fraction of athermal ω phase, is a reversible function of temperature whose value at any temperature, say T_i, is

$$f_\omega^{T_i} = (\chi_\beta - \chi)/(\chi_\beta - \chi_\omega) \ , \qquad (2\text{-}7)$$

according to Eqn. (2-5). An application of this analysis to the results of a susceptibility temperature dependence investigation of a series of Ti-V alloys has enabled $\Delta f_\omega|_{150\text{K}}^{300} \equiv f_\omega^{150\text{K}} - f_\omega^{300\text{K}}$ to be calculated and plotted versus V concentration as in *Fig. 2-7*. The quantity $\Delta f_\omega|_{150\text{K}}^{300}$ is the mole-fraction of ω phase which appears and disappears reversibly as the temperature is cycled between 300 K and 150 K [COL78].

2.3.3 Magnetic Studies of Isothermal ω-Phase Precipitation

During the isothermal aging of a Ti-TM alloy, within the metastable $\omega + \beta$-phase regime, the magnetic response to the approach to $\omega + \beta$ meta-equilibrium can be described by means of the following equation, derived from Eqn. (2-5) and similar to Eqn. (2-6):

$$\Delta\chi = -\underbrace{\frac{\Delta N}{N}\left[\left(\frac{\partial\chi}{\partial c}\right)_\omega - \left(\frac{\partial\chi}{\partial c}\right)_\beta\right]}_{(a)} + \underbrace{(\chi_\omega - \chi_\beta)\Delta f_\omega}_{(b)} \,,$$

(2-8)

where ΔN (>1) represents the number of moles of solute which are transferred from ω to β during the aging of N moles of alloy. As before, the first term, (a), represents an intrinsic effect — this time, the difference between the susceptibility composition dependences of the ω and β phases. The second term, (b), represents the susceptibility change in response to an allotropic change in the alloy's structure between ω and β. Recognizing that χ_ω is always less than χ_β, Eqn. (2-8) shows that, if the composition dependences $\chi_\omega(c)$ and $\chi_\beta(c)$ are exactly parallel, the susceptibility change with aging (invariably a decrease, [Ho73][Col75^b]) will be a direct result of the allotropic $\beta \to \omega$ transformation component, (b); otherwise, this decrease will be aggravated if $(\partial\chi/\partial c)_\omega$ is more positive than $(\partial\chi/\partial c)_\beta$ (as it turned out to be in Ti-V) or partially offset if the converse is true. By exploiting these principles it has, for example, been possible to obtain magnetic estimates of the responses of quenched Ti-V(15 at.%) and Ti-V(19 at.%) to aging at 300°C [Col75^b]. A typical result is given in *Fig. 2-8*, in which a comparison has been made with the results of the more direct measurements of HICKMAN [Hic68, Hic69^a].

2.3.4 Magnetic Studies of Phase Equilibria

In Ti-SM alloys, and particularly the Ti-Al system quenched from various temperatures, magnetic susceptibility procedures have assisted in the investigation of phase equilibria [Yao61][Col70^a, Col79, Col82]. Particular attention has been devoted to Ti-Al in which single-phase-disordered (α) and long-range-ordered (α_2 and γ) regions alternate with two-phase fields whose boundaries can then be determined by the "tie-line" method [Yao61].

Although susceptibility-composition characteristics of single-phase alloys are generally curvilinear, any line crossing a two-phase field must be uncompromisingly straight, as in *Fig. 2-9*. That this is so can be demonstrated by combining Eqn. (2-5) and its compositional counterpart:

$$c = f_\omega c_\omega + (1 - f_\omega)c_\beta$$

(2-9)

in such a way that:

$$\chi = \frac{c_B\chi_A - c_A\chi_B}{c_B - c_A} + \frac{\chi_B - \chi_A}{c_B - c_A}\cdot c \,,$$

(2-10)

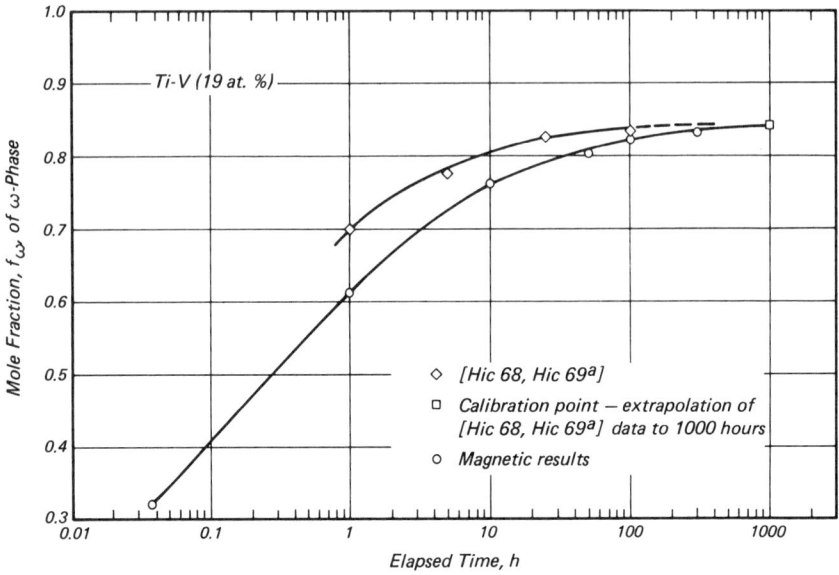

Fig. 2-8. Magnetic study of 300°C-aging-induced ω-phase precipitation in a Ti-V alloy. The results are in good accord with those of HICKMAN, from whose work the calibration point, $f_\omega(1000h/300°C) = 0.84$, was taken [COL75[b]].

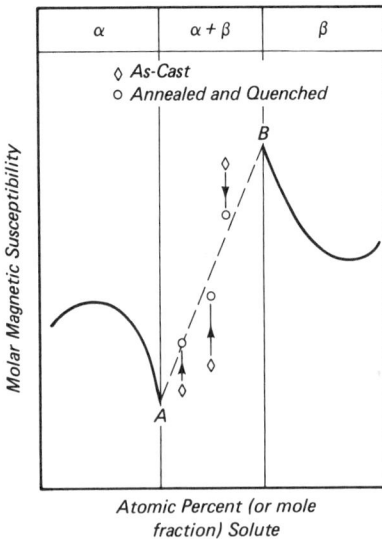

Fig. 2-9. Tie-line concept in the determination of equilibrium phase boundaries. The method requires well-defined "single-phase curves" (insensitive to annealing temperature), in the construction of which some extrapolation may be necessary near the phase boundaries. The concentration dependence of χ in the two-phase region is linear and is constructed either through datum points or on their "far sides" (with respect to some reference condition — say, as-cast) [COL79].

which implies that a plot of χ versus c for a series of equilibrated two-phase alloys is indeed linear with intercept $(c_B \chi_A - c_A \chi_B)/(c_B - c_A)$ and slope $(\chi_B - \chi_A)/(c_B - c_A)$. This is the "tie-line", proper identification of whose endpoints can result in the accurate determination of a pair of phase boundaries. In practice, several series of alloys are prepared, equilibrated at a set of temperatures, and quenched. The quenched structure is assumed to reflect that at equilibrium, due regard being given to the possibility of athermal transformation (such as $\beta \rightarrow \alpha'$ or $\alpha \rightarrow \alpha_2$ for Ti-Al alloys) which, however, does not influence the position of the tie-line endpoints. From the family of magnetic "isothermals" so generated, loci of endpoints can be constructed to form the equilibrium phase boundaries. The results of applying this technique to a determination of the portion of the equilibrium phase diagram for Ti-Al within the composition range 30 to 57 at.% Al and the temperature range 900-1315°C are presented in Chapter 3 (see *Fig. 3-6*).

2.3.5 Magnetic Studies of Texture

In hexagonal-close-packed crystals, magnetic susceptibility, as with other second-rank tensor properties, may be assigned two principal components $\chi_\parallel$ and $\chi_\perp$ directed parallel and perpendicular, respectively, to the hexad axis. It follows that an average susceptibility, $\chi_{av} = \frac{1}{3}\chi_\parallel + \frac{2}{3}\chi_\perp$, may be obtained as the result of a single measurement of an ideal polycrystalline sample. But the large number of randomly oriented grains required may not be present in a small as-cast specimen. Grain size may, of course, be reduced to microscopic dimensions by cold work followed by recrystallization, but then randomness of orientation cannot be guaranteed. Deformation generally induces texturization, which may survive, or even be enhanced by, subsequent heat treatments. Provided proper precautions are followed, however, it is still possible to obtain a χ_{av} from measurements on a textured specimen. In addition, if $\chi_\parallel$ and $\chi_\perp$ values are available from measurements on a single crystal, it is possible to take advantage of the above effect by employing magnetic susceptibility to make quantitative estimates of *bulk* (as distinct from surface) texturization.

Thus with α-phase Ti-SM alloys, which are magnetically anisotropic, an opportunity exists for using magnetic susceptibility techniques in the study of basal-pole texture. The technique recommended, now referred to as the "double-rotation method", was developed by COLLINGS and SMITH in 1968 [Col68] for the determination of the monocrystalline principal susceptibility components $\chi_\parallel$ (parallel to the c-axis) and $\chi_\perp$ (within the basal plane) of hcp crystals.

(a) **Determination of $\chi_\parallel$ and $\chi_\perp$.** The measurement of $\chi_\parallel$ and $\chi_\perp$ is described with reference to *Fig. 2-10* and *Fig. 2-11*. An arbitrary reference plane is ground on a single-crystal specimen of unrecorded orientation; it is then suspended, in turn, along each of the two directions which are orthogonal to the reference direction (along which the susceptibility is χ_{ref}) and rotated through angles θ' and θ''. The resulting susceptibility oscillations possess a common turning point, χ_{common}, which by geometry is always $\chi_\perp$. If when θ' and θ'' are equal to 90° the corresponding susceptibilities are χ' and χ'', respectively, then since

$$3\chi_{av} = \chi_{ref} + \chi' + \chi'' \ , \tag{2-11}$$

and

$$3\chi_{av} = \chi_\parallel + 2\chi_\perp = \chi_\parallel + 2\chi_{common} \ , \tag{2-12}$$

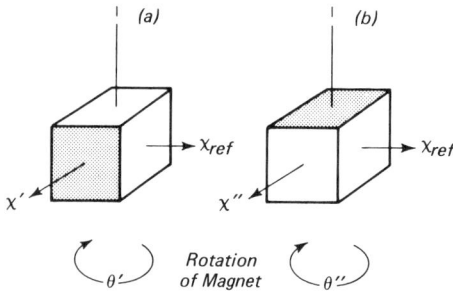

Fig. 2-10. "Double-rotation method" for the determination of three mutually orthogonal magnetic susceptibility components. The average susceptibility is, of course, $\chi_{av} = (\chi_{ref} + \chi' + \chi'')/3$ [COL82].

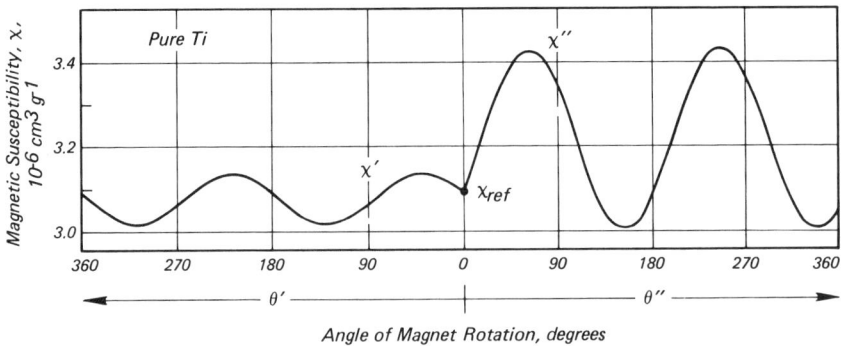

Fig. 2-11. Magnetic susceptibility of a sample suspended from a vertical fiber versus the angle (θ) of an applied magnetic field rotating about the sample-suspension as axis. The common minimum is $\chi_\perp$; $\chi_\parallel$ is then given by $3\chi_{av} - 2\chi_\perp$, where χ_{av} is given (as before) by $(\chi_{ref} + \chi' + \chi'')/3$ [COL80].

enough information is available with which to compute the remaining unknown, $\chi_\parallel$.

The double-rotation technique may also be applied to a textured sample regarded as a "pseudocrystal" characterized by new pseudo-principal susceptibility components whose magnitudes are functions of the monocrystalline $\chi_\parallel$ and $\chi_\perp$ and the degree of texturization.

(b) Determination of Texture. The simplest texture models are those in which the basal poles are distributed symmetrically about some preferred direction. If the susceptibility in that direction is designated $\chi_\parallel'$, and that in the plane normal to it by $\chi_\perp'$, then it can be shown that

$$\chi_\parallel' = \chi_\parallel - \Delta\chi \qquad \qquad (2\text{-}13a)$$

and

$$\chi_\perp' = \chi_\perp + \Delta\chi/2 \; , \qquad \qquad (2\text{-}13b)$$

where $\Delta\chi$ is a measure of the magnetic anistropy introduced by the texture [COL82]. Within this context, two model distribution functions have been considered.

(i) *A rectangular (or step) distribution function* in which all basal pole directions lying within a cone of semi-vertical angle ϕ_c are equally probable. In this case:

$$\chi_\parallel' = \chi_\parallel - 2A(1-Q)/3 \; , \qquad \qquad (2\text{-}14a)$$

and

$$\chi_\perp' = \chi_\perp + A(1-Q)/3 \; , \qquad \qquad (2\text{-}14b)$$

where

$$A \equiv \chi_\parallel - \chi_\perp \; , \qquad \qquad (2\text{-}15a)$$

and

$$2Q \equiv cos\phi_c(1 + cos\phi_c) \; , \qquad \qquad (2\text{-}15b)$$

and where, of course, $2A(1-Q)/3$ plays the role of the $\Delta\chi$ of Eqn. (2-13).

(ii) *A cosine distribution function* between $\phi = 0$ and $\pi/2$, ϕ being an angle that some direction makes with the preferred direction, in which case:

$$\chi'_\parallel = \chi_\parallel - A/2 \tag{2-16a}$$

and

$$\chi'_\perp = \chi_\perp + A/4 \ , \tag{2-16b}$$

which are much simpler functions of the anistropy, A, than those described in Eqns. (2-15).

(c) **Development of a Texture Parameter.** Double-rotation experiments similar to that depicted by *Fig. 2-10* and *Fig. 2-11* serve to determine first $\chi'_\perp$ (the common minimum) and then $\chi'_\parallel$. A single rotation experiment could of course yield $\chi'_\perp$ immediately, and after insertion in Eqn. (2-14) yield a value for the texture-parameter, Q (or ϕ_c), provided single-crystal data were available. This would, however, involve a comparison of $\chi'_\perp$ with $\chi_\perp$ (for the single crystal) determined in a separate experiment, and expose the result to uncertainties arising from positioning and other errors inherent in absolute susceptibility determination. These difficulties can be completely avoided by working in terms of magnetic *anisotropies*.

Full double-rotation measurements yield:

$$A = \chi_\parallel - \chi_\perp \ , \qquad \text{(monocrystal)} \tag{2-17a}$$

and

$$A' = \chi'_\parallel - \chi'_\perp \ . \qquad \text{(textured sample)} \tag{2-17b}$$

These in turn yield the texture-parameter, Q, which according to Eqns. (2-14) is none other than A'/A. The alternative texture index, ϕ_c, if needed, can then be obtained by solving Eqn. (2-15b).

Both methods are fully described in [COL82]. A set of results for a series of cold-rolled Ti-Al alloys is given in *Table 2-1*.

2.4 Low-Temperature Specific Heat

As indicated in Sect. 2.1.4, the specific heat, C, of a normal metal at low temperatures (below $6 \sim 10$ K) can be expressed as the sum of an electronic component, $C_e = \gamma T$, and a lattice component, βT^3. Clearly C/T, when plotted versus T^2, is linear with intercept, γ, and slope β. In case the sample is a superconductor, however, the electronic specific heat acquires an additional component, C_{es}, at the transition temperature, T_c, such that according to BCS theory [BAR57]:

Table 2-1. Texturization Parameters ("Isotropic Model") for Cold-Rolled Ti-Al Alloys [COL82]

Al concentration, at.%	Reduction in thickness by cold rolling, %	Magnetic susceptibility components, χ, 10^{-6} cm^3 g^{-1}			Texturization parameters	
		χ_{av} $= (\chi_{\parallel} + 2\chi_{\perp})/3$	A $= (\chi_{\parallel} - \chi_{\perp})^*$	A' $= 3(\chi_{av} - \chi'_{\perp})^{**}$	Q $= A'/A$	ϕ_c, degrees [from Eqn. (2-15b)]
0.0	25	3.16_6	0.51_5	0.14_4	0.28_0	66
	50	3.17_3		0.29_7	0.57_7	48
3.2	25	3.11_7	0.41_5	0.08_4	0.20_2	72
	50	3.11_6		0.29_4	0.70_8	38
5.5	25	3.11_6	0.35_3	0.21_0	0.59_5	46
	50	3.11_3		0.25_5	0.72_2	37
10.6	24	3.09_8	0.23_3	0.11_7	0.50_2	52

*From monocrystalline results.

**From textured polycrystalline results, see [COL82] for further details.

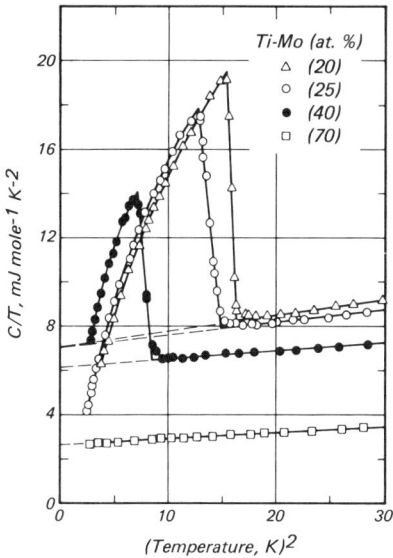

Fig. 2-12. Low-temperature specific heat results for quenched Ti-Mo(20-70 at.%) alloys plotted in the usual format C/T versus T^2. The sharp jumps in the specific heat take place at the superconducting transition temperatures [Col70[a], Col71[d], Col72[a]][Ho73[a]].

$$C_{es}\big|_{T_c} = 2.43\,\gamma T\big|_{T_c} \ , \qquad (2\text{-}18a)$$

or

$$\frac{\Delta C}{\gamma T}\bigg|_{T_c} = 1.43 \ . \qquad (2\text{-}18b)$$

Thus as the sample temperature decreases, a sharp jump in specific heat takes place as soon as the transition temperature is encountered (see *Fig. 2-12*). The position of the jump gives, of course, the transition temperature, T_c, while its relative height, $\Delta C/\gamma T_c$, when compared with 1.43, yields a measure of the degree of "completeness" of the transition.*

2.4.1 The Debye Temperature, θ_D

The lattice specific heat coefficient, β, if expressed in the units J mole^{-1} K^{-4}, yields a low-temperature value of the Debye temperature via the formula:

$$\theta_D = \left(\frac{1.944 \times 10^3}{\beta}\right)^{1/3} \quad (K) \ . \qquad (2\text{-}19)$$

---

*i.e., a measure of how much of the sample is actually participating in the transition, particularly in the case of two-phase material.

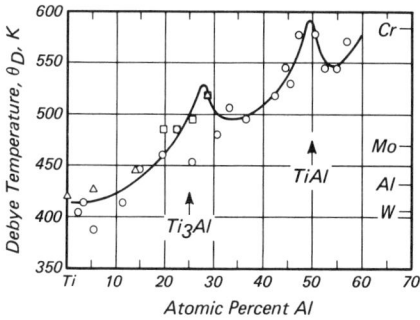

Fig. 2-13. Calorimetrically measured Debye temperature, θ_D, for Ti-Al alloys. *Condition*: as-cast (o); ordered (□); various other heat treatments (△). The Debye temperatures of several pure metals are inserted for comparison [COL80, COL82[a]].

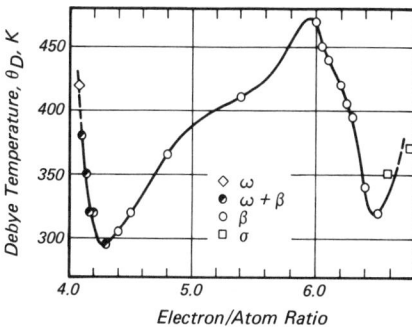

Fig. 2-14. Debye temperature as a function of electron/atom ratio for binary alloys of the Ti-Mo-Re sequence. Particularly noteworthy are that: (i) an $e/a = 6$ guarantees a maximum in the stiffness of the bcc alloys; (ii) at sufficiently low e/a's, the occurrence of ω phase begins to stiffen the bcc lattice; (iii) at sufficiently high e/a's, the lattice is stiffened by a transformation to σ phase [COL73].

θ_D may be regarded as a kind of bulk stiffness modulus. It is well known that the directional interatomic bonding favored by the majority of intermetallic compounds [COL71[b]] is associated with elastic stiffness, hardness maxima, and brittleness. Thus, it is not surprising to find in Ti-Al, a typical Ti-SM system, local maxima in θ_D corresponding to the positions of the brittle intermetallic compounds Ti_3Al and TiAl, *Fig. 2-13*.

Turning now to Ti-TM alloys, a comparable set of studies has also been undertaken on the prototype β-isomorphous system Ti-Mo. *Fig. 2-14*, which displays the calorimetrically measured θ_D as a function of electron/atom ratio for a series of quenched alloys, shows: (i) a continuous softening of the bcc lattice with decreasing Mo concentration; (ii) then with further decrease of Mo concentration, a pronounced stiffening of the lattice due to the appearance of ω-phase precipitation, the occurrence of which is clearly related to the lattice-softening effect just referred to [COL72, COL74].

2.4.2 The Superconducting Transition Temperature: Response to Aging

Low-temperature specific heat makes a useful tool for the monitoring of aging in Ti-TM superconductors. Studies have been conducted on Ti-Fe(7.5 at.%) aged for 1170 h at 175°C followed by an additional 88 h at

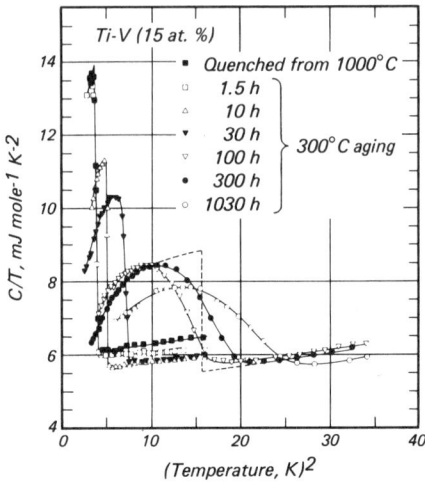

Fig. 2-15. Effects of prolonged aging at 300°C on the low-temperature specific heat of Ti-V(15 at.%). The relative height of the specific-heat jump for specimens with broad transitions may be graphically estimated by extrapolating data above and below the superconductive transition to a vertical line positioned at the transition mid-point (see, for example, the dashed line for the 300-h aging result). As aging proceeds, the transition broadens, the jump-height decreases, but T_c increases [COL75^c].

300°C [Ho73]; Ti-Mo(10 at.%) for 880 h at 350°C [COL72^a][Ho73^a]; and Ti-V(15 and 19 at.%) for 1030 and 2200 h, respectively, at 300°C [COL75^c]. In Ti-Fe, as with Ti-Mo, T_c decreases with aging time and the transition remains fairly sharp, observations which are consistent with the maintenance of a complete proximity effect (precipitate radius < coherence length*) during the development of ω-phase precipitation. Just the opposite is true for the Ti-V alloys; according to *Fig. 2-15*, for example, the maximum T_c *increases* while the transition broadens and the volume fraction of superconducting phase (as gauged by the usual $\Delta C / \gamma T_c$ criterion) decreases. These facts can be explained in terms of a growth of precipitate size (radii becoming $\gg$ coherence length) accompanied by a solute enrichment of the β phase (hence an increase in its T_c) for the alloy compositions concerned [COL75^c].

2.4.3 The Superconducting Transition Temperature: Response to Deformation

This subject has not been investigated extensively. Among the few studies which have been made of the influence of deformation on the superconducting transition were the resistive measurements of Sn by SWANSON and QUENNEVILLE [SWA73] and the calorimetric investigations of Nb by ZUBECK et al. [ZUB79]. Of particular interest in this context, however, are the results of the measurements of Ho and COLLINGS of several plastically deformed Ti-TM alloys.

---

*The characteristic size of the superconducting quasipartical (electron-pair).

In alloys of Ti with 4.5 at.% Mo [Col70], 5 at.% Mo [Ho71][Col71^c], and 7 at.% Mo [Ho71], it has been noted that T_c is raised as a result of deformation-induced-martensitic or twinning transformations; likewise, the addition of 1 or 3 at.% Al to Ti-Mo(5 at.%), which again influences martensitic transformation, results in an increase in T_c [Col76]. Following an earlier suggestion by STRONGIN et al. [STR68], the observed T_c-enhancement was initially attributed to a mechanism which required localized soft-phonon modes to be associated with displaced atoms in the deformed structure [Col70^b]. More recently, however, as a result of the computer fitting of an "asymmetrical-Gaussian-distributed" BCS-specific-heat function to the experimental calorimetric data in the vicinity of the transition, Fig. 2-16, it has been possible to advance a somewhat more plausible argument couched in metallurgical terms [Col78^a]. For example, the specific heat results for Ti-Mo(5 at.%), in which the deformation raises T_c from 1 K to about 3 K (Fig. 2-16) can be interpreted in terms of 68% α'' martensite with $T_c = 3.27$ K, 32% ($\omega + \beta$) with $T_c = 1.0$ K, plus a proximity effect between the two. In Ti-Mo(7 at.%), in which twinning is believed to be the primary deformation product, the results of the fitting exercise can be interpreted in terms of 63% original $\omega + \beta$ phase, 27% low-T_c twin-boundary and highly defected material, and 10% high-T_c ω-deficient twin-boundary phase.

2.4.4 The Superconducting Transition Temperature: Low-Concentration Quenched-Martensitic Ti-TM Alloys

Superconducting transitions associated with the quenched martensitic ($\alpha^m \equiv \alpha'$ or α'') structure have been investigated calorimetrically in several systems, notably Ti-V and Ti-Nb [Hei64], Ti-Mn and Ti-Co [Hak64], Ti-Fe [Bat64], and Ti-Mo [Col69]. Comparative studies of the superconducting transition in an extensive series of dilute Ti-TM alloys

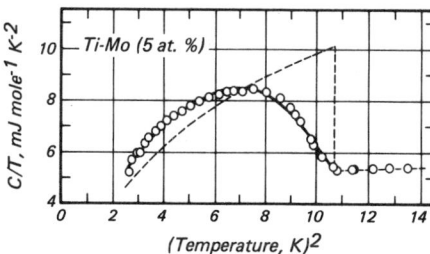

Fig. 2-16. Low-temperature specific heat results for quenched-plus-deformed Ti-Mo(5 at.%) plotted in the usual format C/T versus T^2 and fitted with a Gaussian-rounded BCS-specific-heat function. Best fit to the data (full line) was achieved with an "extreme negative" skew distribution (i.e., left-half Gaussian, or $f = -1.0$*). The unrounded function is shown as a broken line [Col78^a], see also [Whi76].

*The symmetrical Gaussian is parameterized by $f = 0.0$, see [Whi76].

have been undertaken by BUCHER *et al.* [BUC65] (Ti-V, -Cr, -Mn, -Fe, -Nb, -Mo), who applied a Gaussian rounding technique to the analysis of the specific heat jump in much the same manner as that referred to above, and subsequently by AGARWAL [AGA74] (Ti-Sc, -V, -Cr, -Mn, -Fe, -Co, -Ni, -Hf). Taken together, the results of both workers lead to the following conclusions: (i) the low-temperature specific heat of α^m-Ti-Mn has a temperature dependence characteristic of the localized-magnetic-moment behavior referred to earlier; (ii) the specific heat jumps in alloys such as Ti-V, -Nb, and -Mo are not unduly rounded; and (iii) those of the alloys Ti-Cr, -Fe, and -Co are exceptionally broad. A subdivision of the alloys into two groups which include (i) Ti-V, -Nb, and -Mo on one hand and (ii) Ti-Fe on the other, with Ti-Cr occupying an intermediate position, is apparent in *Fig. 2-17.*

For a rationalization of the above-mentioned behavior we turn again to a metallurgical explanation, this time in terms of the relative rates of diffusion of transition-element solutes in β-Ti. *Fig. 2-18*, which intercompares tracer diffusivities [ZWI74, p. 108], indicates that of the alloys for which data are available, only Ti-V, -Nb, and -Mo have the opportunity to transform athermally to α^m during quenching from the β phase, while in alloys such as Ti-Fe, -Co, and -Ni, whose diffusivities are almost two orders of magnitude higher, significant levels of solute redistribution and Widmanstätten growth can be expected during the quenching of the moderately massive samples needed for the usual kind of low-temperature specific heat measurement.

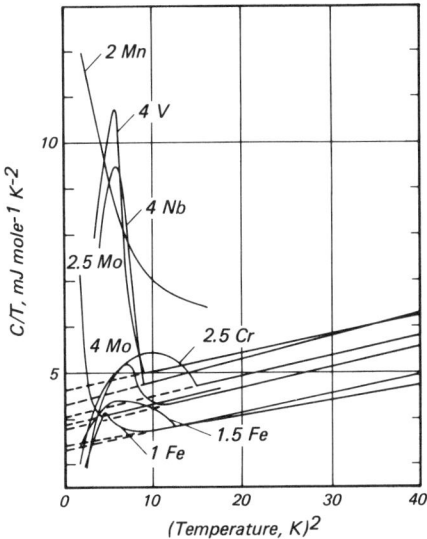

Fig. 2-17. Low-temperature specific heats in the vicinity of their superconducting transitions for low-concentration martensitic Ti-V, Ti-Nb, Ti-Mo, Ti-Cr, Ti-Mn, and Ti-Fe alloys, indicating a decrease in the abruptness of the specific heat jump on proceeding from the "β-isomorphous" to the "β-eutectoid" class of alloys [BUC65].

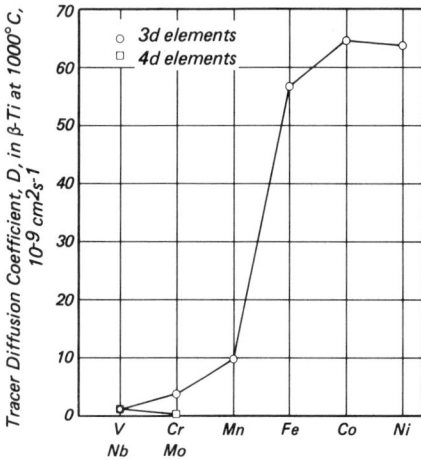

Fig. 2-18. Tracer diffusion coefficients for the 3*d* solutes V, Cr, Mn, Fe, Co, and Ni, and the 4*d* solutes Nb and Mo, in β-Ti at 1000°C — computed from frequency-factor and activation-energy data of ZWICKER [ZwI74, p. 174].

2.5 AC Impedance Measurement

LUHMAN and colleagues, using an AC inductive technique, have measured the position, width, and fine structure associated with the superconducting/normal transition in Ti-Cr alloys as part of an extensive study of precipitational effects associated with quenching, aging, and "upquenching" [LUH69, LUH70, LUH70ª, LUH71]. As indicated in Sect. 2.1.5, impedance (inductance) changes in a sample/coil system were detected by measuring the voltage drop across a coil (surrounding the alloy sample) supplied with 1-kHz current from an oscillator. The experimental results were displayed as plots of "impedance" versus sample temperature [LUH70ª] or the first-derivative of inductance, with respect to temperature, versus temperature [LUH69]. The manner in which the impedance results are interpretable can be described with the aid of *Fig. 2-19* for as-quenched and quenched-plus-aged (51min/300°C) Ti-Cr(15 at.%). In part *(a)* of the figure, the asymmetric shape of the transition curve was ascribed to composition gradients in the all-bcc alloy. During aging at 300°C, athermal ω-phase precipitation was supposed to take place. The existence of ω phase after 51min/300°C was responsible for the double superconducting transition barely detectable in part *(b)* of the figure. But double transitions are more easily detected and resolved in the first-derivatives of the impedance-temperature curves. Some results for a Ti-Cr(10.3 at.%) alloy in the as-quenched and quenched-plus-aged (28min/196°C) conditions are given in *Figs. 2-20(a)* and *(b)*, respectively. Peak *A* was interpreted as being due to ω phase. The double peak *B,C* was identifiable with the β phase, since it generally occurred independently of the presence of ω

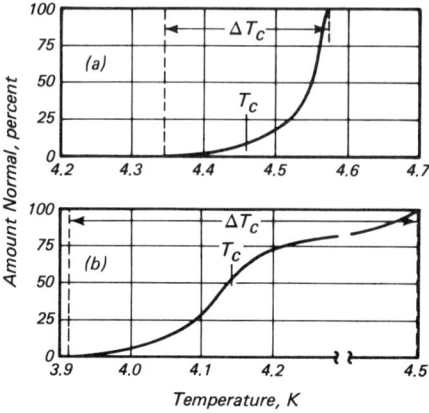

Fig. 2-19. Variation of impedance with temperature of a sample of Ti-Cr(15 at.%) in two metallurgical conditions: *(a)* solution treated 1h/850°C/WQ; *(b)* solution treated plus aged 51min/300°C [Luʜ70, p. 104; Luʜ70[a]].

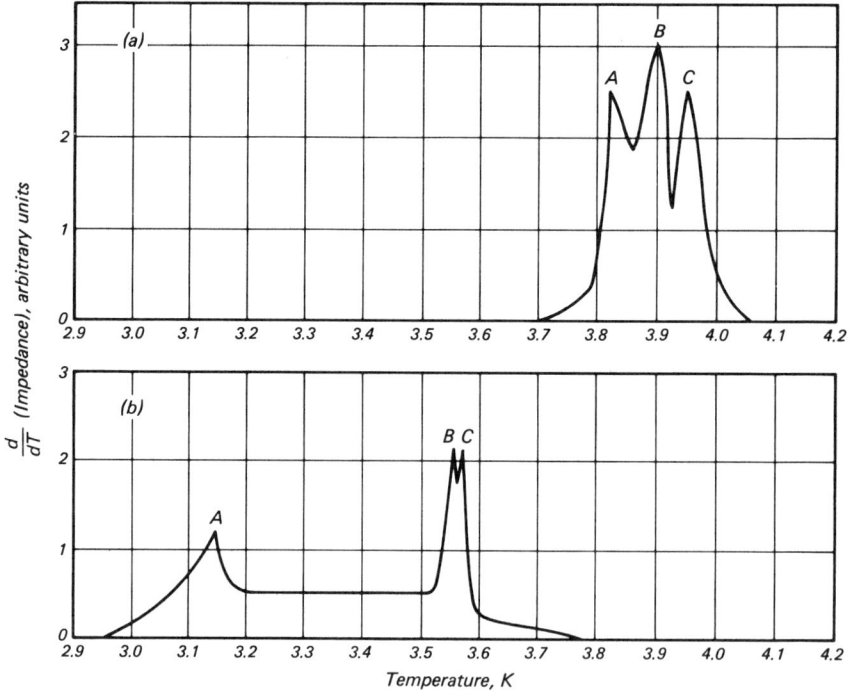

Fig. 2-20. First derivative of impedance with respect to temperature versus temperature for a sample of Ti-Cr(10.3 at.%) in two metallurgical conditions: *(a)* solution treated $1\frac{1}{2}$h/980°C/WQ; *(b)* solution treated plus aged 28min/196°C [Luʜ69].

phase. The doublet nature of the peak was taken as an indication of the presence of solute gradients.

A particularly interesting phenomenon characteristic of $\omega + \beta$-phase Ti-TM alloys is ω-phase reversion. The impedance-measurement technique has been used to study this effect in Ti-Cr(9.5 at.%), Ti-Cr(15 at.%), and Ti-V(24.4 at.%) [LUH70, LUH71]. Following an aging heat treatment at 300°C to produce isothermal ω phase, both alloys were "up-quenched" to 450°C, where they were held for 3 min prior to water quenching. The up-quenching of aged Ti-Cr(15 at.%) raised its T_c from 3.05 K to 4.292 K, the latter being less than the 4.456 K of the as-quenched alloy. The up-quenching of Ti-V(24.4 at.%) yielded a T_c of 5.089 K, higher than the 4.382 K of the as-quenched sample. The observed differences in T_c between the as-quenched, $\omega + \beta$-aged, and $\beta' + \beta$-reverted samples were interpreted in terms of composition fluctuations, screening of precipitation zones by the surrounding matrix, the effects of coherency strain fields, and the T_c-composition dependences of the individual components.

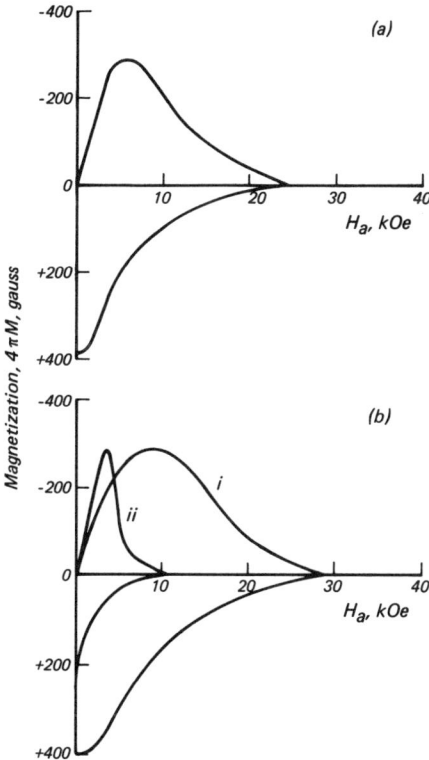

Fig. 2-21. Magnetization versus applied field curves (see *Fig. 2-1*) for a sample of Ti-V(28 at.%) in three metallurgical conditions: *(a)* solution treated 1h/1000°C/WQ plus aged 10h/350°C; *(b) (i)* solution treated plus aged 24h/350°C, *(ii)* that followed by 3min/550°C [LUH70, Sect. V(K); LUH72].

2.6 Magnetization of Superconducting Titanium-Transition-Metal Alloys

Magnetization measurements, which yield values of the upper critical field, H_{c2}, and generally exhibit irreversibility, provide additional indirect information on precipitation and solute redistribution. LUHMAN and colleagues [LUH70, LUH72] have also employed this technique in studies of ω-phase precipitation and aging, ω-phase reversion, and α-phase precipitation in some representative Ti-V and Ti-Nb alloys. Some typical results are depicted in *Fig. 2-21*. In *Fig. 2-21(a)*, representing Ti-V(28 at.%) aged 10h/350°C, the magnetic irreversibility is due to ω-phase precipitation. This is enhanced by extending the heat treatment to 24 h at the same temperature, *Fig. 2-21(b)*, curve *(i)*, and diminished following a heat treatment of 3min/500°C (followed of course by quenching) which reverts the ω phase to β, curve *(ii)*. The hysteresis which remains is a consequence of flux-pinning (field trapping) by the modulated $\beta' + \beta$ structure.

Equilibrium Phases

3.1 Titanium Alloy Phases

3.1.1 Systematics of Phase Stability

The question of lattice stability plays an important role in any discussion of the physics of pure metal or alloy systems. This is particularly true of Ti alloys whose lattice stability (i.e., structural phase stability) has technical as well as fundamental significance. The crystal structures of the three long periods of transition elements change more or less systematically from hcp through fcc as the group number increases from IV to VIII. Whether or not there is an underlying physical significance to this, in the case of transition metals a useful correlation certainly exists between crystal structure and group number (in the case of elements) or crystal structure and average group number or electron/atom ratio (in the case of alloys). The existence of such correlations suggests that electronic structure plays an important role in the control of phase stability.

Numerous workers have attempted to define the factors which govern the existences of the α and β phases of Ti alloys. As indicated in Chapter 1, solute atoms which lower the temperature of the allotropic $\alpha \rightarrow \beta$ transformation, with respect to that of pure Ti, are referred to as β stabilizers. Conversely, α stabilizers raise that temperature. As pointed out by McQuillan [McQ63], the relatively more open bcc structure has a higher vibrational entropy than do the close-packed structures hcp and fcc. Consequently, during heating, the free energy of an imaginary bcc lattice will decrease more rapidly than those of the competing alternatives such that eventually a temperature will be reached whereat the lattice (if it does not melt) will transform from the low-temperature-stable close-packed structure (generally hcp, α) to bcc. Underlying this thermodynamic picture is an atomistic model involving electronic cohesive forces (directional or otherwise) and atomic size effects. Jaffee, in an early analysis of the situation [Jaf58], suggested that atomic size effect was the dominant factor; subsequently, he was able to conclude that, although size effect needed to be taken into consideration, the dominant phase-stabilizing mechanism was electronic in nature. MCQUILLAN also took this latter view [McQ63], but pointed out that exceptions did of course exist—for example, the β-stabilizing tendencies of the solutes Bi and Pb were thought to be due to their relatively large atomic sizes [McQ63].

Factors controlling the stabilization of the α and β phases in Ti alloys have also been discussed in several publications by COLLINGS and GEGEL [Col73[a], Col73[b], Col75[a]], with particular reference to the Ti-Al and Ti-Mo systems. Stability was qualitatively discussed both from electronic [Col73, Col82[a]] and thermodynamic (phenomenological) [Col75[a]] standpoints.

(a) Electronic Considerations. As a result of low-temperature specific heat measurements, it was noted that the more stable of a pair of allotropes was associated with the lower electronic density-of-states at the Fermi level, $n(E_F)$. This rule was exemplified using data for the following pairs of competing phases: α_2* and α; α and β; ω and β [Col73].

(b) Thermodynamic Considerations. Pair-interaction-potential calculations based on the relative-vapor-pressure measurements by HOCH *et al.* [Rol71, Rol72] have divided the field of Ti-base alloys into two regimes: (i) β-stabilized Ti-TM alloys whose regular-solution thermodynamic interaction parameter, Ω_{ij}, is positive (indicative of clustering systems), and (ii) α-stabilized Ti-SM alloys for which Ω_{ij} is negative (short-range-ordering systems) [Col75[a]].

All authors agree that the alloys of Ti can be assigned to one of two major categories**—α-stabilized or β-stabilized systems. MARGOLIN [Mar60] has recommended subdividing the former into two more groups according to the degree of α stabilization: (i) those of "limited α stability", in which decomposition of α takes place by peritectoid reaction into β plus a compound (e.g., Ti-B, Ti-C, and Ti-Al), and (ii) those of "complete α stability", in which the α phase can coexist with the liquid (e.g., Ti-O and Ti-N). MARGOLIN has also recommended subdividing β-stabilized alloys into four categories in the following way: (i) β-isomorphous systems such as Ti-Mo and Ti-Ta which show restricted α- and extensive β-solubility ranges; (ii) β-and-α-isomorphous systems, such as Ti-Zr, showing complete mutual solubilities in both the α and β phases; and (iii) β-eutectoid systems in which the β phase has a limited solubility range and is able to decompose into α and a compound (e.g., Ti-Cr and Ti-Cu)—this class being further subdivisible into two more depending on whether the β decomposition

---

*A hexagonal DO_{19} structure found in the Ti-Al system.

**As indicated in Sect. 1.2, technical alloys are, of course, classified as α, β, and $\alpha+\beta$ according to their microstructural states when placed in service.

is rapid (e.g., Ti-Cu, Ti-Ni, and Ti-Sn) or sluggish (e.g., Ti-Cr, Ti-Mn, and Ti-Fe). KORNILOV [KOR82] has discussed a subdivision into what he refers to as four basic alloy types. Although the classes were untitled, their descriptions conformed to the categories: α phase, β-and-α isomorphous, β isomorphous, and β eutectoid, referred to above. In a survey of Ti alloy phases, MOLCHANOVA [MOL65, p. xiv] has offered a detailed subdivision of the equilibrium phase diagrams into two groups of three sub-categories, and one group of four. The schematic phase diagrams which typify these ten sub-categories, and the solutes which give rise to each of them, are depicted in *Fig. 3-1*. The descriptions of the groupings are as follows:

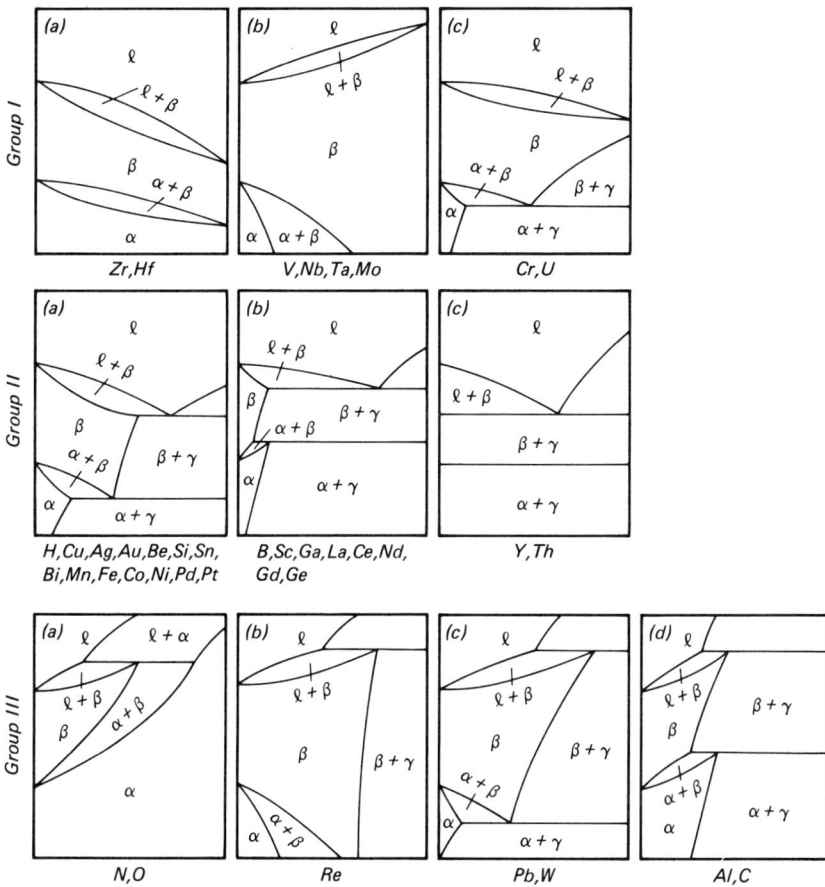

Fig. 3-1. Classification scheme for binary Ti-alloy phase diagrams [MOL65, p. xiv].

Group I: Systems with continuous β solid solubility
   I(a) Complete miscibility in the α phase
   I(b) Partial miscibility in the α phase
   I(c) Partial miscibility in the α phase and eutectoid decomposition of the β phase
Group II: Eutectic systems
   II(a) Partial miscibility in the α and β phases; eutectoid decomposition of the β phase
   II(b) Partial miscibility in the α and β phases; peritectic decomposition of the β phase
   II(c) No detectable solid solubility
Group III: Peritectic systems
   III(a) Simple peritectic
   III(b) Partial miscibility in the α and β phases
   III(c) Partial miscibility in the α and β phases; eutectoid decomposition of the β phase
   III(d) Partial miscibility in the α and β phases; peritectic dissociations of the β phase

In the same book, MOLCHANOVA has also offered a simpler subdivision into the four categories depicted in *Fig. 3-2* [MOL65, p. 154].

3.1.2 Electronic Factors in the Stability of Phases, Particularly in Transition Metals

With transition metals, the electron/atom ratio, e/a, is the same as the average "group number" — referring to the numbers assigned to the groups of the periodic table. Thus, e/a takes on the values 4 through 10 when applied to the members of the seven columns of the transition-metal (TM) block of the periodic table headed by the elements Ti through Ni. The e/a is a parameter in terms of which numerous physical and mechanical properties of binary TM alloys, particularly Ti-TM, can be conveniently displayed. Several important physical (including superconductive) properties may also be indexed in terms of quantities related to the above-mentioned conventional e/a, viz: the atomic-volume-corrected "electron concentration" of JENSEN, MATTHIAS, and ANDRES [JEN65] or the "effective electron/atom ratio", N_{eff}, of DESORBO et al. [DES65]. Another quantity advocated by LUKE, TAGGART, and POLONIS [LUK64] as being appropriate for the indexing of the compositional threshold for martensitic transformation in Ti-TM alloys is an average Pauling valence which, although equal to conventional e/a for the groups IV through VI transition elements, never exceeds the value 6 for elements of later groups. The crystal structures, particularly those of simple metals, have been justified from

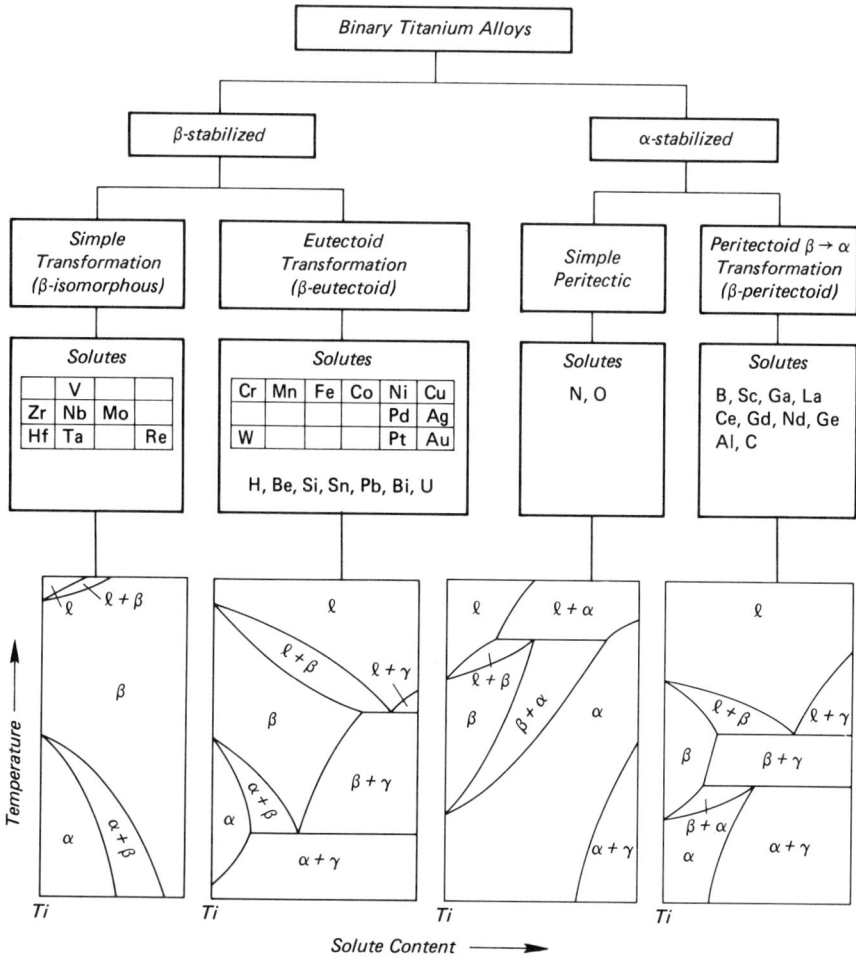

Fig. 3-2. Classification scheme for binary Ti-alloy phase diagrams—an alternative to the previous figure. α and β are solid-solution alloys; γ represents an intermetallic compound [MOL65, p. 154].

several fundamental standpoints. BREWER and ENGEL [BRE67] have related structure to the spectroscopic states of the individual participating atoms. PAULING [PAU67], in considering the metallic bond, has also used this as a basis for discussion. The OPW type of approach also utilized atomic spectroscopic states, but in a more satisfactory manner by starting with an array of bare ions and then replacing the electrons in such a way that their wavefunctions represent tightly bound electrons near the cores, and nearly free electrons in the spaces between. Although attempts to deal electronically with phase stability in transition metals have been made by INGLESFIELD [ING69] and PETTIFOR [PET72], the situation with regard to alloys is much more difficult.

Very successful calculations of the *electronic structures* of alloys, and in particular the manner in which the band density of states, $n(E)$, varied with energy, E, have been made using the coherent potential approximation (CPA) first applied by EHRENREICH and colleagues [KIR70] to the Cu-Ni system. The particular method used, since it took a tight-binding (TB) approach to the d-electrons and a nearly-free-electron (NFE) one to the other electrons in the band, has been referred to as the NFE-TB-CPA. Although it was especially applicable to Cu-Ni, it was the forerunner of more sophisticated methods, developed by others, of dealing with the energy-band structures of disordered alloys [FAU82]. In overcoming the limitations of the NFE-TB-CPA, a CPA-method was developed which had some features in common with the old Korringa-Kohn-Rostoker (KKR) method. The first publication of a full KKR-CPA calculation, again as it applied to Cu-Ni alloys, was by STOCKS, TEMMERMAN, and GYORFFY [STO78]. The number of alloy systems to which such calculations have been applied, and the results compared with experiment (angular resolved photoemission is a favored method), has been quite limited.

But it is still a large step from calculations of this kind to calculations of lattice phase (crystal structure) stability. PETTIFOR [PET79] has made considerable progress toward the calculations of the heats of formation of binary alloys by using a simple formalism, based on a Friedel expression for the binding-energy per atom, in which the CPA played a fundamental role. As indicated above, it is a remarkable experimental fact that the crystal structures of $3d$, $4d$, and $5d$ transition metals, and their "adjacent" binary alloys, vary in a regular manner from hcp through bcc to fcc as a function of the e/a or average group number. MOTT and JONES' interpretation of one of the Hume-Rothery rules was an unsuccessful attempt to provide a crystal-structure/electron-concentration relationship for nontransition metals; other approaches have been more successful [BLA67]. So far the empirical crystal-structure ("phase stability") versus e/a relationships as they apply to *transition metals* seem to exist without a general theoretical interpretation [FAU82, p. 186].

The closest approach to an exact calculation of phase stability in a transition-metal alloy system, in particular Zr-Nb, has been made by MYRON, FREEMAN, and MOSS [MYR75], who dealt not with equilibrium phases but with an electronic mechanism leading to the appearance of the metastable ω phase. Adequately discussed in their paper (see also SINHA and HARMON [SIN76]), the technique employed coupled a KKR band-structure and Fermi-surface calculation for bcc Zr with the effect of "rigid-band" modifications of it brought about by the addition of Nb, in order to demonstrate that electronically instigated enhancement of the natural dip in the bcc-lattice phonon spectrum at $\frac{2}{3}\langle 111 \rangle$ could lead, in a manner to be discussed below, to the ω-phase transformation. Purely electronic descriptions of equilibrium-phase stability have been strongly criticized from two standpoints by KAUFMAN and NESOR [KAU73]. They noted that: (i) in many treatments, competition between phases was completely ignored; and (ii) when electronic property data acquired at low temperatures were used to justify high-temperature phase transformations, no account was taken of the entropy differences. KAUFMAN and NESOR recommended the use of a thermodynamic procedure, in which the energetic competition between candidate phases was fully taken into account, when attempting to define the lattice stabilities of metallic elements as well as alloy systems. Full discussion of a quantitative thermodynamic approach, leading to the computer-assisted calculation of binary and multicomponent phase diagrams, is to be found in the work of KAUFMAN [KAU70].

For the remainder of this section we abandon predictions and return to a phenomenological electronic justification for the existences of the two principal classes of Ti alloys: the α and the β phases.

3.1.3 α-Titanium Alloys

α-stabilizing solutes are those which, as function of concentration, more or less elevate the temperature of the $\beta/(\alpha+\beta)$ transus. Such solutes are generally non-transition metals (i.e., "simple metals", SM). An explanation of α stability based on electron-screening arguments proceeds as follows: When simple metals (e.g., Al) are dissolved in Ti, very few electrons appear at the Fermi level, most of them going to states within the lower part of the band. The Ti d-electrons tend to avoid the Al atoms, which thereby have the effect of diluting the Ti sublattice. The consequence of this is to emphasize any pre-existing Ti-Ti bond directionality and thus to preserve the hcp structure characteristic of the Ti crystal. In general, when simple metals are added to Ti, the fields of Ti-like α stability are eventually terminated by intermetallic compounds, of composition Ti$_3$SM,* which are also hexagonal in structure. The bond argument is consistent with the observation that α stabilizers are quite rapid solution strengtheners either in hcp solid solution or when added to bcc alloys

---

*An exception is supposed to be Ti$_4$Pb.

[GEG73[a]]. The classification of α-phase alloys into systems whose phase diagrams exhibit (i) peritectic transformations or (ii) peritectoid transformations, according to MOLCHANOVA's simplified scheme, is shown in *Fig. 3-2.*

3.1.4 β-Titanium Alloys

The transition-metal block of the periodic table may be regarded as commencing with group III, Sc, Y, and La (or perhaps more precisely, Lu). In this scheme, the alkaline-earth metals, Ca, Sr, and Ba, may be regarded as "pre-transition metals", and the noble metals, Cu, Ag, and Au, as "post-transition metals". As indicated in most periodic charts of the elements, the structures of the transition metals all change from hcp to bcc as e/a increases from 4 through 6. It is possible that stabilization of the bcc structure can be justified within the framework of a screening model in terms of which a high conduction-electron concentration, which enhances the screening of ion cores, may favor a symmetrical, hence cubic, structure. Thus an increase in electron density (as in the groups V and VI elements), which tends to symmetrize the screening, increases the stability of the bcc structure. Symmetrization may also be accomplished through lattice vibrations; thus, all six of the groups III and IV elements transform to the bcc structure at high temperatures (as compared to their Debye temperatures). With regard to alloys, the addition of transition elements to Ti increases the electron density and consequently stabilizes the bcc or β structure. Thus, as a general rule, the transition elements are "β stabilizers". The systematics of β stabilization by transition elements has been discussed in detail by AGEEV and PETROVA [AGE70], according to whom: (i) the β-stabilizing action of TM solutes is greater the "farther" they are from Ti in the periodic table; and (ii) for the retention of the metastable-β solid solution during quenching, the β stabilizer has to provide for an electron/atom ratio of at least 4.2.

According to ZENER [ZEN48], and subsequently FISHER [FIS70, FIS75], who has considered the problem of bcc stability in considerable detail, the magnitude of the elastic shear modulus $C' = (C_{11} - C_{12})/2$ is a useful parameter for ranking the stabilities of bcc transition metals and alloys. The variation of C' with conventional electron/atom ratio is plotted in *Fig. 3-3*, which shows that the alloying of group-IV elements with other elements to the "right" of them in the periodic table increases the bcc stability, which rises to a maximum near $e/a = 6$ for the elements Cr, Mo, and W. On the other hand, with decreasing e/a, the vanishing of C' for $e/a = 4.1$ corresponds to the compositional threshold for martensitic transformation—to be discussed below. As a result of the alloying of Ti with transition elements of higher group number, the continuous increase of bcc stability manifests itself as a lowering of the $\beta/(\alpha+\beta)$ transus temperature. As indicated already in *Fig. 3-2*, within the context of β stabilization two subclasses of phase diagram exist – the "β isomorphous" and the "β eutectoid", depending on whether or not a solid-solution/compound eutectoid exists at a sufficiently elevated temperature. It would be instructive in the present context to consider a group of simplified, compositionally truncated, binary Ti-TM equilibrium phase diagrams, arranged according to the positions that the solute elements occupy in the TM block of the periodic table. Some representative diagrams selected from such a postulated arrangement are presented in *Fig. 3-4*. In order to focus attention on the alloys of most interest, the limiting composition (in at.%) in each group (except group IV itself) has been selected such that $e/a \le 5.0$. In so doing it has been assumed that the number of $s + d$ valence electrons belonging to the elements in the columns headed by Fe, Co, Ni, and Cu is 8, 9, 10, and 11, respectively. An alternative way of deriving a reduced composition scale for intercomparison purposes, and one that would focus attention on alloy chemistry rather than electron density, might have been to normalize composition (i.e., stretch the composition scale) to that of the first β-eutectoidal intermetallic compound.

Interesting systematics to be noted in *Fig. 3-4* are that: (i) as the solute element moves to the "right", the phase diagram changes from the β-isomorphous to the β-eutectoidal type; and (ii) along the row Mn-Fe-Co-Ni-Cu, the eutectoid temperature increases monotonically. Extrapolating this trend to the "left" suggests that Ti-V can also be thought of as eutectoidal, but with an inaccessibly low eutectoid temperature.

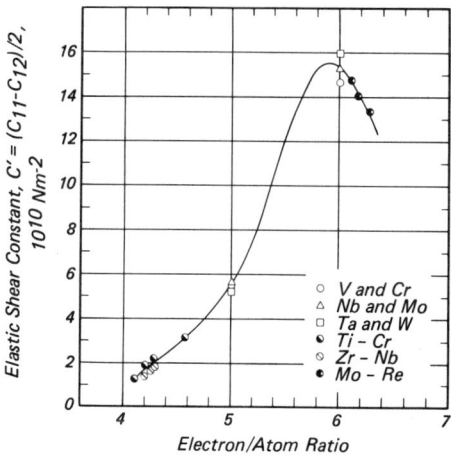

Fig. 3-3. Elastic shear modulus $C' = (C_{11} - C_{12})/2$ for bcc binary transition-metal alloys as a function of electron/atom ratio [COL73[a]].

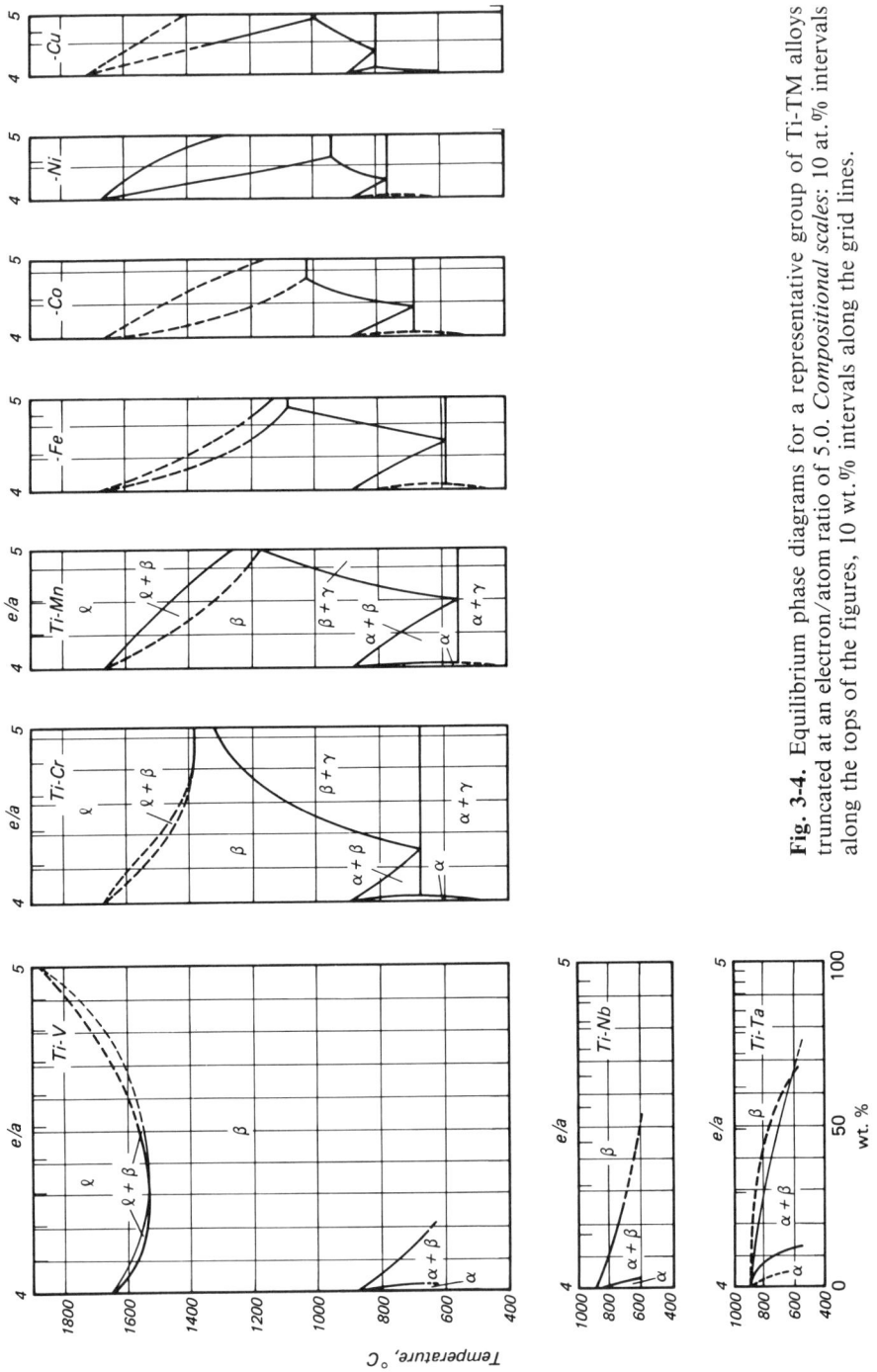

Fig. 3-4. Equilibrium phase diagrams for a representative group of Ti-TM alloys truncated at an electron/atom ratio of 5.0. *Compositional scales*: 10 at.% intervals along the tops of the figures, 10 wt.% intervals along the grid lines.

3.2 Binary Titanium Alloys

The equilibrium phase diagrams of numerous Ti-base binary alloys have been presented and discussed by McQUILLAN and McQUILLAN [McQ56ᵃ], IMGRAM *et al.* [IMG61], and ZWICKER [ZWI74], and the properties of several important systems have been reviewed by JAFFEE [JAF58] and MARGOLIN and NIELSON [MAR60]. The most comprehensive compendium of binary phase diagrams has of course been provided by MOLCHANOVA [MOL65].

In what follows, four alloy systems will be briefly reviewed. They are: Ti-Al, the basis of technical α-Ti alloys; Ti-Mo, a β-isomorphous alloy, and the basis of several technical β-Ti alloys; Ti-Nb, another β-isomorphous alloy, of importance as a superconductor; and Ti-Cr, a β-eutectoid alloy.

3.2.1 A Representative α-Stabilized Titanium Alloy: Titanium-Aluminum

Equilibrium phase diagrams of this representative α-stabilized Ti-base binary alloy are to be found in [McQ56ᵃ, p. 174], [MOL65, p. 137], and [ZWI74, p. 147]. Research articles in which descriptions of portions of the equilibrium diagram have been discussed are listed in *Table 3-1*. Two of the most interesting and important regions of the diagram surround the ordered intermetallic compounds Ti_3Al and $TiAl$. Short-range and long-range order in Ti-Al alloys and the occurrence of the long-range-ordered α_2 phase in the vicinity of the compound Ti_3Al was discussed by BLACKBURN [BLA67ᵃ], who offered an equilibrium partial phase diagram for the composition range 0-30 at.%, *Fig. 3-5*. Confirmatory evidence for the existences and ranges of the disordered and ordered phases in Ti-Al (20 ~ 30 at.%) is to be found in the results of the magnetic susceptibility measurements of COLLINGS *et al.* [COL75ᵃ, p. 152]. GEHLEN [GEH70] has investigated the crystallography of Ti_3Al, which was found to possess the DO_{19} structure—a unit cell composed of four regular hcp cells apparently supported by covalent-like directional bonds connecting the Al and Ti atoms. Solution strengthening in Ti-Al alloys in terms of localized Ti-Al bonds, leading at sufficiently high solute concentrations to the above-mentioned long-range-ordered structure and electronic effects related to it, have been discussed by COLLINGS and co-workers [COL70ᵃ, COL75ᵃ, COL82ᵃ]. Some physical properties of Ti-Al alloys attributable to the occurrence of brittle intermetallic compounds at compositions near 25 at.% (Ti_3Al), 50 at.% ($TiAl$), and ~70 at.% (~$TiAl_3$) have been discussed, with particular reference to electronic bonding and its relationship to intermetallic compound formation, by COLLINGS [COL82ᵃ].

Table 3-1. List of Investigations Directed Toward a

Al concentration range, at.%	Temperature range for equilibrated solid alloys, °C	Principal and auxiliary techniques described
0-64	750-1100	Optical metallography; with x-ray diffraction and thermal analysis
0-75	700-1400	Optical metallography; with x-ray diffraction and Vickers hardness
0-75	700-1200	Optical metallography; with x-ray diffraction, Vickers hardness, thermal analysis and centrifugal bend tests
5-49	550-1050	Electrical resistivity, magnetic susceptibility; with optical metallography and x-ray diffraction
0-63	450-1350	Electrical resistivity; with optical metallography and x-ray diffraction
0-48	800-1450	Optical metallography and x-ray diffraction
5-38	400-1100	Magnetic susceptibility
0-38	550-1200	Optical metallography, electrical resistivity and x-ray diffraction
5-43	550-1200	Electrical resistivity and Vickers hardness; with thermal analysis, dilatometry, and x-ray diffraction
7-35	550-1100	Optical metallography; with electron microscopy, x-ray diffraction, differential thermal analysis, electrical resistivity and dilatometry
5-25	500-1100	Electron microscopy
27-45	1025-1225	Electron microscopy
30-57	900-1365	Magnetic susceptibility; with optical metallography

The wide discrepancies that exist among the numerous Ti-Al phase diagrams presently in existence is evident in ZWICKER's collection of six equilibrium diagrams [ZWI74, p. 147]. In order to shed further light on the position of a particularly important feature, the $(\alpha_2 + \gamma)/\gamma$ phase boundary, COLLINGS employed magnetic susceptibility techniques (augmented by optical metallography) in order to develop an equilibrium partial diagram for Ti-Al(30-57 at.%) within the temperature range 900 ~ 1300°C [COL79]. The results of the magnetic work (already described in Sect. 2.3.4) are presented in *Fig. 3-6*; some related optical micrographs are presented in *Fig. 3-7*. The single γ phase is clearly evident in the micrographs for Ti-Al(51, 52 at.%); however, a precipitate of as-yet unknown origin always occurred in Ti-Al(55 at.%) quenched from temperatures below about 1300°C.

Determination of the Ti-Al Equilibrium Phase Diagram

References

OGDEN, H.R., MAYKUTH, D.J., FINLAY, W.L., and JAFFEE, R.I., *Trans. TMS-AIME* **191**, 1190 (1951).

BUMPS, E.S., KESSLER, H.D., and HANSEN, M., *Trans. TMS-AIME* **194**, 609 (1952).

KORNILOV, I.I., PYLAEVA, E.N., and VOLKOVA, M.A., *Izv. Akad. Nauk SSSR, Otd. Khim. Nauk* **No. 7**, 771 (1956).

SAGEL, K., SCHULTZ, E., and ZWICKER, U., *Z. Metallkund.* **46**, 529 (1956).

SATO, T. and HUANG, Y., *Trans. Jpn. Inst. Metals* **1**, 22 (1960).

ENCE, E. and MARGOLIN, H., *Trans. TMS-AIME* **221**, 151 (1961).

YAO, Y.L., *Trans. ASM* **54**, 241 (1961).

CLARK, D., JEPSON, K.S., and LEWIS, G.I., *J. Inst. Metals* **91**, 197 (1962).

KORNILOV, I.I., PYLAEVA, E.N., VOLKOVA, M.A., KRIPYAKEVICH, P.I., and MARKIV, V.Ya., *Dokl. Akad. Nauk, SSSR* **161**, 842 (1965).

CROSSLEY, F.A., *Trans. TMS-AIME* **236**, 1174 (1966).

BLACKBURN, M.J., *Trans. TMS-AIME* **239**, 1200 (1967).

BLACKBURN, M.J., *The Science, Technology and Application of Titanium*, Pergamon Press, 1970, p. 633.

COLLINGS, E.W., *Met. Trans.* **10A**, 463 (1979).

3.2.2 Representative β-Isomorphous Alloy Systems: Titanium-Molybdenum and Titanium-Niobium

The β-isomorphous alloy systems Ti-Mo and Ti-Nb have been deliberately selected — Ti-Mo as a prototype β-stabilized technical alloy, of interest from a mechanical-property standpoint, Ti-Nb as an important superconducting alloy.

(a) **The Titanium-Molybdenum System.** Equilibrium phase diagrams for Ti-Mo have been developed by CRAIGHEAD and co-workers (1950),* HANSEN and co-workers (1951),* DUWEZ (1951),* MOLCHANOVA [MOL65, pp.

---

*See references in [MOL65, p. 32].

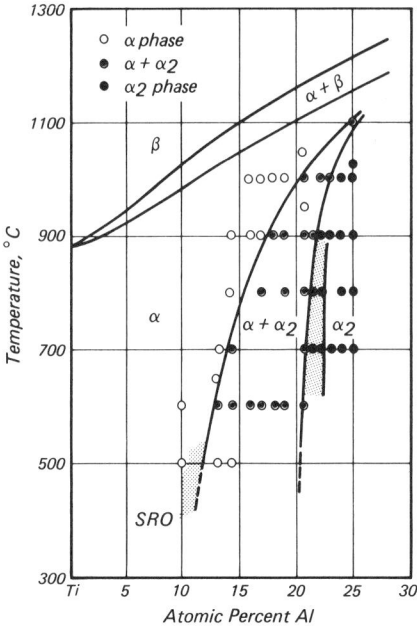

Fig. 3-5. Partial Ti-Al equilibrium phase diagram for the range 0-25 at.% Al [Bla67[a]].

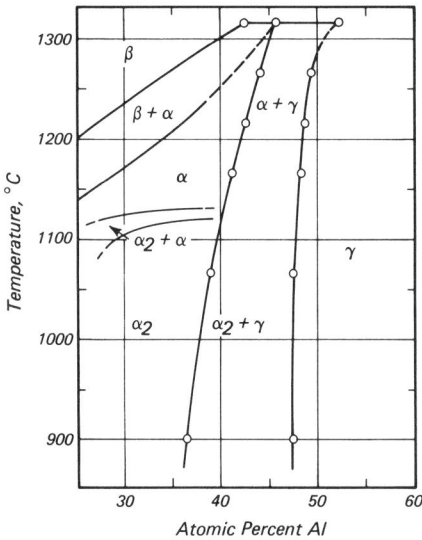

Fig. 3-6. Partial Ti-Al equilibrium phase diagram for the range 25-57 at.% Al. The data points (o) were recently determined magnetically (see Sect. 2.3.4 and [Col79]); the boundaries of the $\beta + \alpha$ and $\alpha_2 + \alpha$ fields were earlier established by BLACKBURN [Bla70].

27-32], and most recently by TERAUCHI et al. [Ter82]. The diagram according to HANSEN et al. [Han58, p. 977] is reproduced in *Fig. 3-8*. The alloys used in developing this diagram had been homogenized (20-40h)/ 1250°C prior to being annealed at eight temperatures between 855 and

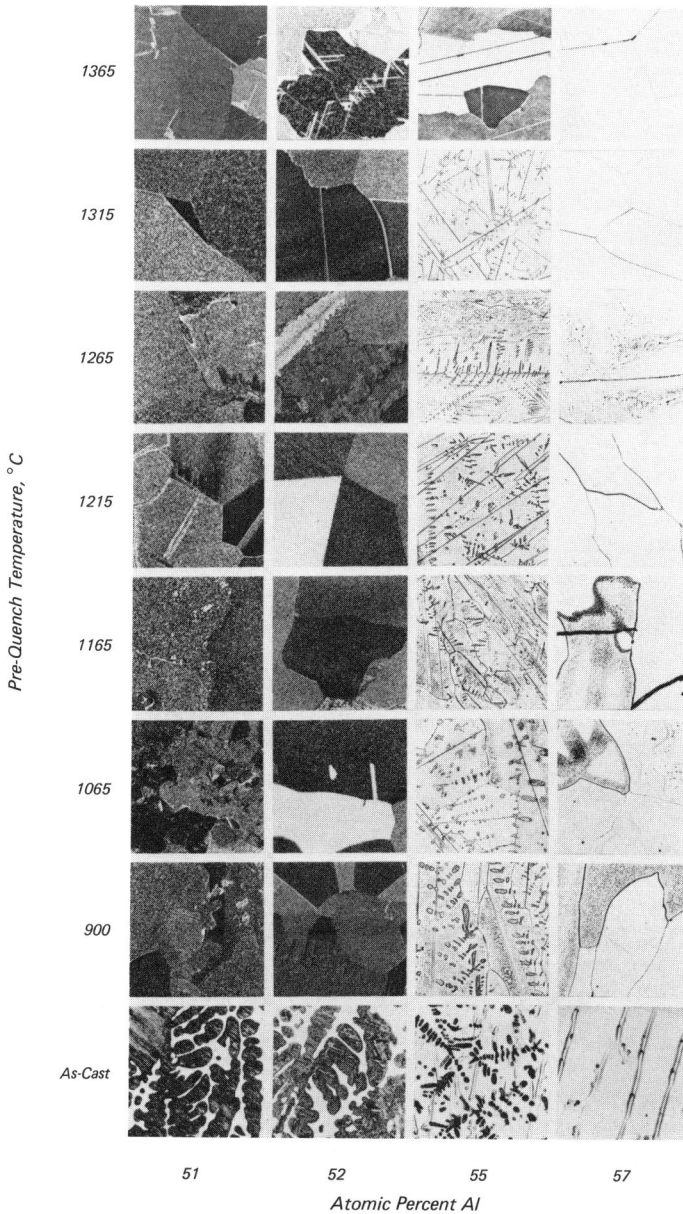

Fig. 3-7. Optical micrographs of the alloys Ti-Al(51 through 57 at.%) quenched into ice brine from equilibration anneals at the temperatures indicated (see [COL79] for details of the heat treatments). Magnification of the original 6 × 6 cm² micrographs, 200× [COL79]. Copyright © 1979, *Metallurgical Transactions*, reprinted with permission.

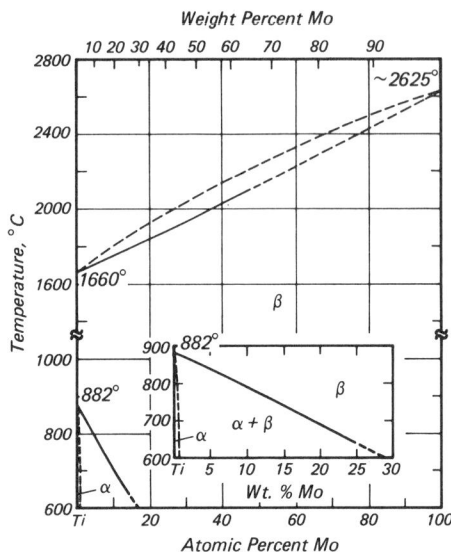

Fig. 3-8. The Ti-Mo equilibrium phase diagram [HAN58, p. 977].

600°C for times of between 90 and 650 h. Diagrams such as *Fig. 3-8* are not usually continued below 600°C owing to the difficulties which are always encountered when attempting to attain thermodynamic equilibrium in reactive alloys when the diffusion rates are slow. According to MOLCHANOVA [MOL65, pp. 27-32], the β phase is stable at all temperatures in alloys containing more than 16 at.% Mo (28 wt.%). The $\beta/(\alpha+\beta)$ phase boundary is almost linear and intersects the 650°C line at 14 at.% Mo (24 wt.%). The maximum solubility of Mo at 600°C in α-Ti-Mo alloys was stated to be only about 0.4 at.% Mo (0.8 wt.%), with the cautionary note that there was some uncertainty associated with that number.

Non-equilibrium quenched phases and the effects of aging on them as they proceed toward thermodynamic equilibrium are to be discussed in subsequent sections of this book. It will be pointed out that within the equilibrium $\alpha+\beta$ field, and bordering the meta-equilibrium $\omega+\beta$ zone, the aging of quenched β-stabilized alloys can result in a separation of the β phase into a solute-rich β matrix and a solute-lean β' precipitate. The β'/β interfaces [LUH70], or the interiors of the β' precipitates themselves [WIL73], are the sites of α-phase precipitation during further aging. Clearly the phase-separated $\beta'+\beta$ is a non-equilibrium condition. A double-bcc phase can, however, exist as an *equilibrium* two-phase state in some alloy system. Referred to as "β-phase immiscibility", it occurs, for example, in Zr-Nb [LOV66] and related systems such as Ti-Zr-Nb [KIT70[a]]. The purpose of this digression into the existences of phase-separated and phase-immiscible double-bcc phases is to provide a suitable context for the discussion

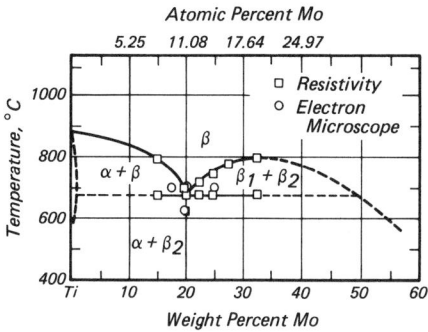

Fig. 3-9. Suggested partial Ti-Mo equilibrium phase diagram due to TERAUCHI *et al.* [TER82].

of the results of some very recent studies of the Ti-Mo system by TERAUCHI and colleagues [TER82]. From optical observations, electrical resistivity measurements, x-ray diffractometry, lattice-parameter measurement, and TEM, those authors have deduced the existence of a pair of bcc phases — referred to as $\beta_1 + \beta_2$ — occupying an area of the equilibrium phase diagram lying outside the $\alpha + \beta$ field, *Fig. 3-9*. Obviously the new diagram will not receive immediate acceptance, and before any modifications to existing diagrams are considered it will be necessary either to re-confirm the validity of *Fig. 3-9,* or to find satisfactory explanations for the observations that gave rise to it.

(b) The Titanium-Niobium System. Since Ti-50Nb (~34 at.% Nb), as well as ternary and quaternary alloys based on it, have found use in the form of Cu-matrix multifilamentary composites in the windings of large superconducting magnets, an accurate knowledge of the equilibrium and meta-equilibrium phase diagrams of the Ti-Nb system is particularly important. A composite equilibrium phase diagram for Ti-Nb, clearly also a β-isomorphous system, is reproduced in *Fig. 3-10* based on the work of HANSEN *et al.* [HAN51] and IMGRAM *et al.* [IMG61] (see also RONAMI *et al.* [RON70], and JEPSON *et al.* [JEP70]). Since thermodynamic $\alpha + \beta$ phase equilibrium is difficult to achieve at temperatures below about 600°C, the experimentally deduced line-diagram is not continued below about that temperature. At temperatures below the $\beta/(\alpha + \beta)$ transus, or its projection, the approach to equilibrium is made via the decomposition of the metastable $\omega + \beta$ or β phases, the rate of which is accelerated in the presence of oxygen or the products of heavy deformation. The recent renewal of interest in the occurrence of α-Ti-Nb precipitation in rather concentrated Ti-Nb alloys of up to about 53.5 wt.% (37 at.%) Nb, and its role in technical superconductivity, has not only rekindled a corresponding level of interest in the equilibrium phase diagram for temperatures in the vicinity

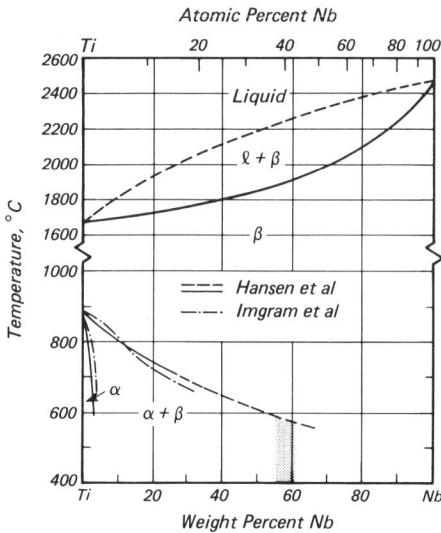

Fig. 3-10. The Ti-Nb equilibrium phase diagram due to HANSEN *et al.* [HAN51] and IMGRAM *et al.* [IMG61] as modified by the observation that no appreciable α-phase precipitation takes place during the aging at 400°C of alloys with more than about 40 at.% Nb [NEA71]—see also [WES82, WES83].

of 400°C, but has also led to the ready availability of heavily cold-worked and heat-treated samples upon which the necessary electron microscopy can be carried out. In this context, HILLMANN *et al.* [PFE68] and WILLBRAND and SCHLUMP [WIL75] have examined the occurrence and morphology of α-phase precipitation in alloys of 50 wt.% Nb after moderate-time aging at temperatures near 380°C, and more recently WEST and LARBALESTIER [WES80, WES82] have observed the presence of α-Ti-Nb precipitates in Ti-53.5Nb after two stages of heat treatment (separated by cold drawing) of 80h/375°C and 40h/375°C. Subsequently, the latter authors conducted a series of high-resolution TEM studies of a number of heavily cold-worked and heat-treated Ti-Nb alloy superconductors, detecting α-phase precipitation in alloys again with Nb concentrations as high as 53.5 wt.% [WES83].

On the other hand, in a well-known study of Ti-58Nb (analyzed composition), heavily deformed but aged usually for relatively short periods of time at temperatures of 350-500°C, NEAL *et al.* [NEA71] were able to detect only *traces* of α-phase precipitation. The failure to observe appreciable precipitation in this case may have been due partly to deficiencies in the detection techniques used and partly to the short aging times. With regard to the latter, the long-time aging of Ti-base alloys is fraught with the danger of oxygen contamination even under the most carefully regulated conditions. Since oxygen is an α stabilizer, its presence can always be interpreted as being partly responsible for the appearances of α precipitates in regions where they might otherwise not be expected. Although the heavy cold work experienced by wires which have been drawn down to small diameters facilitates the attainment of thermodynamic equilibrium in alloys aged at only moderate temperatures, it also makes the detection and identification of any resulting precipitates all the more difficult.

Precipitation in cold-worked superconductive Ti-Nb(36 at.%) alloys has been carefully studied by OSAMURA *et al.* [OSA80]. Specimens were in the form of: (i) *foils* — solution treated 1h/800°C/slow cooled, cold-rolled to thickness reductions of up to 90% and aged at 380°C; and (ii) *fine wires* — obtained from Cu-clad composites which had experienced reductions of 90 to 99.994%, aged at 380°C. Upon the foil samples, small-angle x-ray scattering (SAXS) experiments were performed in order to determine the average diameter of the precipitated particles (twice R_G, the Guinier radius) and their interparticle spacing (derived from R_G assuming a close-packed arrangement of particles). TEM observations were then performed in order to obtain visual images and further information about the particles and their distributions. With regard to the wire samples, although SAXS measurements could be performed using bundles of them, direct observation of the precipitates by TEM was very difficult. Instead, wide-angle x-ray diffraction served to confirm that the dominant precipitate formed during aging was in fact α phase.

In discussing precipitate detection techniques for use in heavily cold worked samples, WEST [WES82] has pointed out that since α-phase precipitates are not easy to identify in such structures, particularly since dark-field imaging is complicated by the close positioning of matrix and precipitate reflections in selected area diffraction (SAD) patterns, the best analytical results are obtained through the use of scanning transmission electron microscopy (STEM) and associated energy-dispersive x-ray analysis (EDAX). The results of such observational methods applied to a sample of Ti-53.5Nb are given in *Fig. 3-11*.

With regard to the compositional range of α-phase precipitation, based on the combined results of the experiments published to date, it is concluded that 56 ± 1 wt.% Nb can be reasonably taken as a practical boundary between the $\alpha + \beta$ and β phases at about 400°C. This result has been inserted in *Fig. 3-10*. In so doing it was recognized that although α-phase precipitation from alloys of concentrations greater than the above limit may be thermodynamically permissible at temperatures below about 400°C, it may not be practically realizable.

3.2.3 A Representative β-Eutectoid Alloy System: Titanium-Chromium

Since Cr is an ingredient of technical alloys such as Ti-13V-11Cr-3Al and Ti-3Al-8V-5Cr-4Mo-4Zr, both metastable-β alloys, an understanding

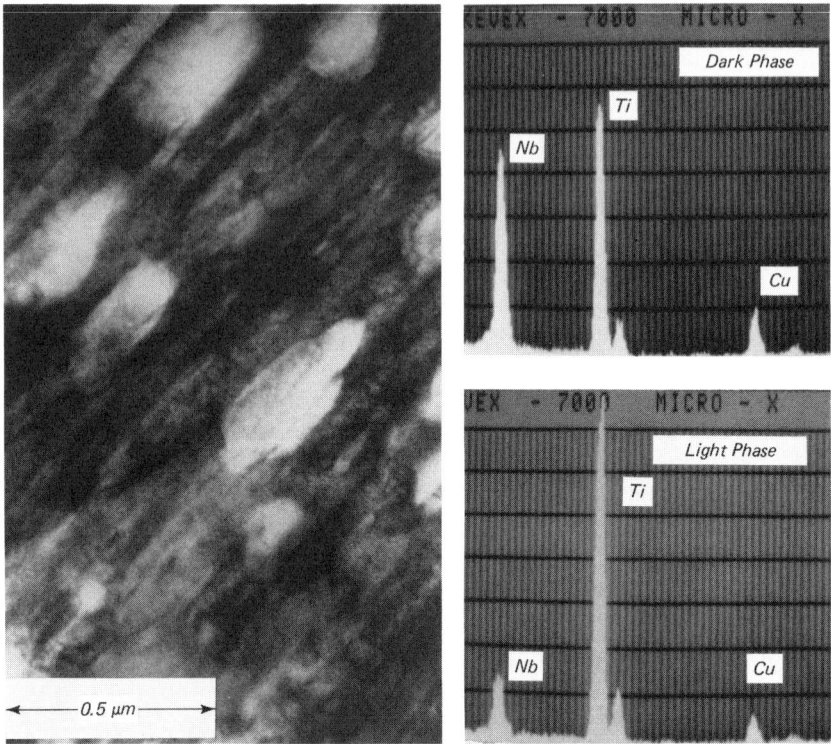

Fig. 3-11. *Left side*: STEM micrograph of a Ti-53.5Nb alloy cold worked to a diameter of 3.66 mm and aged 80h/375°C, then cold worked to a diameter of 1.5 mm and aged 40h/375°C [WES82]. *Right side*: EDAX pictures of the "dark" (Nb-rich, matrix) and "light" (Nb-lean, α-phase precipitate) regions. Photographs courtesy of A.W. West (University of Wisconsin).

of its binary phase diagram with Ti is particularly important. The complete equilibrium phase diagram for Ti-Cr, a β-eutectoid system, is reproduced in *Fig. 3-12*. The source of that figure is HANSEN [HAN58, p. 566]; other standard reference sources such as McQUILLAN and Mc-QUILLAN [McQ56, p. 193] and MOLCHANOVA [MOL65, p. 33] have offered qualitatively similar diagrams, differing from *Fig. 3-12* only in minor details. Of particular interest in systems of this type is the tendency for the Ti-rich bcc phase to decompose eutectoidally into a weak α-phase solid solution plus a compound. In the Ti-Cr system depicted here, Ti-Cr(14 at.%) decomposes very sluggishly at temperatures below $550 \sim 685°C$ (675°C is the value preferred by HANSEN [HAN58, p. 566] and SHUNK [SHU69, p. 282]). The presence of the interstitial elements H, N, and O increases the rate of eutectoid decomposition [MOL65, p. 34]. At higher

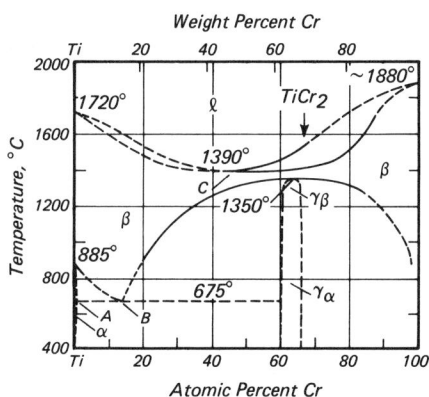

Fig. 3-12. The Ti-Cr equilibrium phase diagram. The points indicated by *A, B,* and *C* are at concentrations of 0.5, 14, and ~45 at.% Cr, respectively [HAN58, p. 566].

and lower Cr levels, hyper- or hypo-eutectoidal decomposition, respectively, can also take place. Once formed, the products of such decomposition are readily re-dissolved during heating in the β field. The intermetallic compound component of the eutectoidal decomposition — represented by the symbol γ in the appropriate diagram of *Fig. 3-2* — is of nominal composition $TiCr_2$ with a "homogeneity range" of some 2 percentage points. Its composition, as function of temperature, is reviewed in [SHU69, p. 283]. $TiCr_2$ is polymorphic, existing as the hcp ($MgZn_2$-structure) "β-$TiCr_2$" phase (labelled γ_β in *Fig. 3-12*) at high temperatures and the fcc ($MgCu_2$-structure) "α-$TiCr_2$" phase (γ_α) at lower temperatures. The transformation temperature of $TiCr_2$ seems to be uncertain [SHU69, p. 283]: according to MOLCHANOVA [MOL65, p. 36], the hexagonal modification exists above 1300°C and the cubic below 1000°C, both phases coexisting in the intervening temperature range. As regards the high-temperature bcc solid solutions, a Ti-rich phase, β', and a Cr-rich, β'', coexist in a temperature-composition zone bounded by 1350-1400°C and 50-70% (wt.% or at.%) Cr [MOL65, p. 34]. The coexistence in thermodynamic equilibrium of β' and β'' is comparable to the $\beta' + \beta''$ immiscibility exhibited by the Zr-Nb system, but should not be confused with $\beta \rightarrow \beta' + \beta$ phase separation, a non-equilibrium state of previously quenched alloys during moderate-temperature aging.

3.3 Multicomponent Titanium-Base Alloys

Having identified a binary alloy with properties more or less suitable for the application in mind, whether it be structural or superconductive, it can generally be improved by the carefully engineered addition of further alloying components. Thus, for example, commencing with Ti-Al, the addition of Sn has led to the technical α alloy Ti-5Al-2.5Sn and the addition of V to the popular $\alpha + \beta$ alloy Ti-6Al-4V. Substitutions of Ta for Nb and/or Zr for Ti have improved the superconductive properties of Ti-50Nb and resulted in technically important ternary and quaternary superconducting alloys. Substitutions of Zr and Sn into the basic β-stabilized Ti-12Mo have yielded the well-known technical alloy, β-III. In structural alloys, the additions are chosen to achieve improvements in mechanical properties such as strength and toughness, structural phase stability, and chemical stability. Some guidelines have already been offered in Sect. 1.4.2.

3.3.1 α Alloys

(a) The Technical α Alloy Ti-5Al-2.5Sn. The total α-stabilizing content, on an at.% basis, in Ti-5Al-2.5Sn is 9.7 at.%. Reference to the binary Ti-Al equilibrium phase diagram, *Fig. 3-5*, suggests that this ternary alloy possesses the highest level of solution strengthening possible while avoiding precipitation of the embrittling α_2 phase. The commercial alloy may, however, contain traces of β phase resulting from contamination by Fe originating in the sponge-Ti used in its preparation [Woo72]. The following microstructures may be developed in Ti-5Al-2.5Sn by appropriate thermomechanical processing: (i) equiaxed-α, obtained by annealing a mechanically worked alloy in the α field (below ~1025°C); (ii) sharp acicular-α, obtained by water quenching from the bcc field (above ~1050°C); and (iii) structures intermediate between these extremes, obtained by furnace cooling from the bcc field and by adjusting the prior-β grain size through appropriate control of the annealing time in that field.

(b) Advanced α-Stabilized Alloys. HOCH *et al.* [Hoc73], drawing an analogy with the Ni-base superalloys and their γ' (Ni_3Al) precipitates, recommended the use of highly alloyed α-phase alloys containing α_2-phase precipitates for high-temperature applications where creep resistance is important. The α_2 phase referred to was understood to be an ordered compound of variable stoichiometry, based on the DO_{19} compound Ti_3SM, where SM may be Al, Ga, In, or Sn.

The binary stoichiometric α_2-Ti_3Al phase is extremely brittle in tension (less so in compression, of course). Accordingly it has been found to severely embrittle the two-phase Ti-Al(>12 at.%) alloys in which it occurs. Some degree of ductility can be acquired if the α_2 particles can be coarsened sufficiently to enable a dislocation bypass (looping) mechanism to operate, but the desired coarsening is difficult to achieve in practice. The goals of high-concentration α-phase alloy development have been to take the greatest possible advantages of solution- and precipitate-strengthening

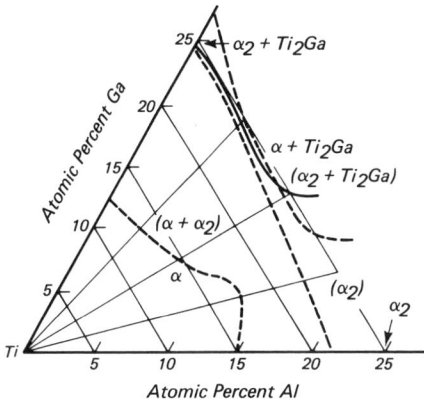

Fig. 3-13. Ti-rich corner of a Ti-Al-Ga equilibrium phase diagram according to SAKAI [SAK69]. The dashed lines and phases in parentheses indicate the range of α_2 proposed by WILLIAMS and BLACKBURN [WIL69] but not observed by SAKAI, see also [HOC73, HOC73[a]].

but at the same time to avoid the previously inevitable α_2-Ti_3Al particle embrittlement. With these goals in mind, considerable effort has been directed toward exploring the microstructural, physical, and mechanical properties of Ti-Al-Ga alloys. The situation has been discussed by GODDEN et al. [GOD73] and HOCH et al. [HOC73] with reference to some earlier relevant studies by BLACKBURN and WILLIAMS [BLA69][WIL69] and LÜTJERING and WEISSMANN [LUT70, LUT70[a]]. An equilibrium phase diagram depicting a corner of the Ti-Al-Ga system is given in *Fig. 3-13*.

With regard to the solution-strengthening aspects, COLLINGS and GEGEL [COL75[a]] have shown that, as functions of total α-stabilizer content, the tensile strengths of Ti-Al$_x$-Ga$_x$ alloys were always greater than those of either Ti-Al$_{2x}$ or Ti-Ga$_{2x}$. The extra strengthening was attributed to secondary solid-solution strengthening arising from Al-Ga interaction. Two alternative approaches to the solving of the α_2-phase embrittlement problem have been discussed: (i) one involved the properties of the matrix and its ability to accommodate to the presence of the precipitate particle; (ii) the other involved the properties of the precipitate particle itself. HOCH et al. [HOC73] tended to focus attention on alternative (i) above. They compared the systems Ti-Al-Sn and Ti-Al-Ga in terms of the parameter of misfit between the α_2 particle and the matrix, and on this basis judged that, whereas dislocation-cutting-type* deformation mechanisms could be expected in the former, a dislocation-bypass deformation mechanism operative in Ti-Al-Ga should have led to at least some measure of ductility. GODDEN and ROBERTS [GOD73] discarded the possible role played by α/α_2 mismatch and suggested that the increased ductility observed in Ti-Al-Ga might have been due to an alloying-induced improvement in the ductility of the α_2 itself. Basing this particular argument on the results of some

---

*i.e., fracture of the precipitate particle.

order-disorder transition temperature (T_{OD}) measurements by COLLINGS *et al.* [COL71[a]], they suggested that with a T_{OD} of ~500°C the α_2 component of an alloy such as Ti_{75}-$Al_{12.5}$-$Ga_{12.5}$, quenched from 1300°C, may be present in the disordered state, and consequently intrinsically more ductile.

3.3.2 $\alpha + \beta$ Alloys

In alloys such as Ti-Al, the two-phase $\alpha + \beta$ region is both *narrow* and *high in temperature*. With Ti-6Al (10 at.% Al), for example, the $\beta/(\alpha + \beta)$ and $(\alpha + \beta)/\alpha$ transformations take place at ~1010°C and 970°C, respectively. The introduction of V at constant Al concentration, although it has a comparatively small influence on the position of the $\beta/(\alpha + \beta)$ transus, produces a rapid decrease in the $(\alpha + \beta)/\alpha$ transus. These effects are shown in *Fig. 3-14*, which plots transformation temperature versus V concentration for two fixed levels of Al, and *Fig. 3-15*, which performs a complementary function in terms of a continuous variation of the Al concentration at four fixed levels of V. The corresponding equilibrium ternary phase diagrams for Ti-Al-V are given in *Fig. 3-16*. Evidently, even in the presence of, say, 10 wt.% Al, that of 8 wt.% V is sufficient to permit the retention of a small β-phase component to temperatures as low as 600°C. The popular alloy Ti-6Al-4V is a member of this system.

(a) The Near-α Alloy Ti-6Al-4V. The alloy Ti-6Al-4V could be regarded as being derived from unalloyed Ti by (i) the addition of Al to produce solution strengthening and raise what becomes the $\beta/(\alpha + \beta)$ transus, and

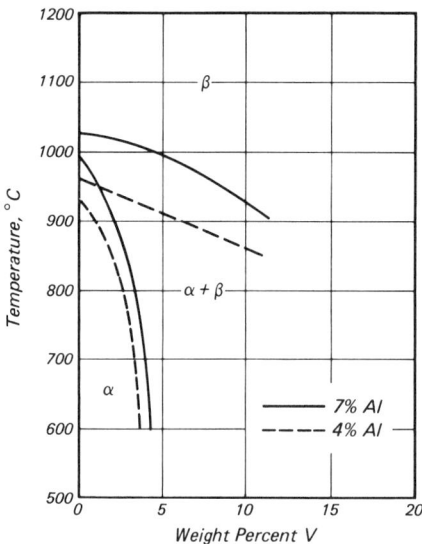

Fig. 3-14. "Vertical" sections of the Ti-Al-V versus T equilibrium-phase solid (right triangular prism) at 4 and 7 wt.% Al [RAU56].

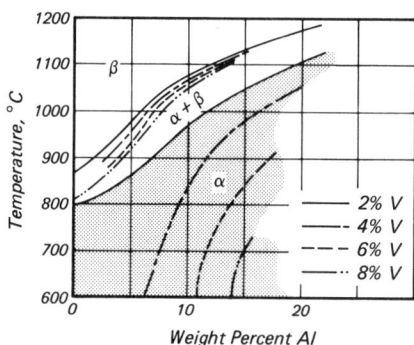

Fig. 3-15. "Vertical" sections of the Ti-Al-V versus T equilibrium-phase solid (right triangular prism) at 2, 4, 6, and 8 wt.% V [RAU56].

(ii) the addition of V to lower the $(\alpha + \beta)/\alpha$ transus. The equilibrium states of this alloy at various temperatures are indicated by special points in *Fig. 3-16*.

A wide range of processing techniques are applied to Ti-6Al-4V to produce numerous kinds of mill products exhibiting a wide range of microstructures. A set of typical microstructures is presented in *Fig. 3-17*.

The existence of a third phase in $\alpha + \beta$ Ti-6Al-4V was first reported by RHODES and WILLIAMS [RHO75]. Designated an "interface phase", it occurred either as a monocrystalline layer with an fcc structure, or as a striated layer about 200 μm thick consisting of platelets. Interface layers have also been reported in Ti-11.6Mo and Ti-14Mo-6Al by MARGOLIN *et al.* [MAR77] and more recently an fcc or fct phase has been observed separating the α and β phases of a β-cooled ($\leqslant 400°C$ per minute) multicomponent alloy based on Ti-6Al-5Zr by HALLAM and HAMMOND [HAL80].

(b) The Near-α Alloy Ti-8Al-1Mo-1V. In the alloy Ti-8Al-1Mo-1V with its relatively large amount of the α stabilizer, Al, the low concentrations of the β stabilizers, Mo and V, which are present permit only small amounts of β phase to become stabilized [WOO72, p. 1-3:72-1]. The $\beta/(\alpha+\beta)$ transus is at approximately 1040°C. On being quenched from within the $\alpha+\beta$ field, from just below the transus down to about 900°C, the β component transforms to α. While being annealed at lower temperatures than these, the β phase becomes sufficiently enriched in Mo and V (see *Fig. 3-14*, again) that it resists transformation during quench. At 870°C and 815°C the alloy contains 16% and 14%, respectively, of the β phase. Although this alloy, like Ti-6Al-4V, must be designated "$\alpha+\beta$, near-α" it does exhibit several α-alloy characteristics such as good weldability and elevated-temperature creep strength. Indeed, it was developed in the first place as a "super-α" alloy* for forged engine components [WOO72].

---

*The term "super" implies suitability for high-temperature service.

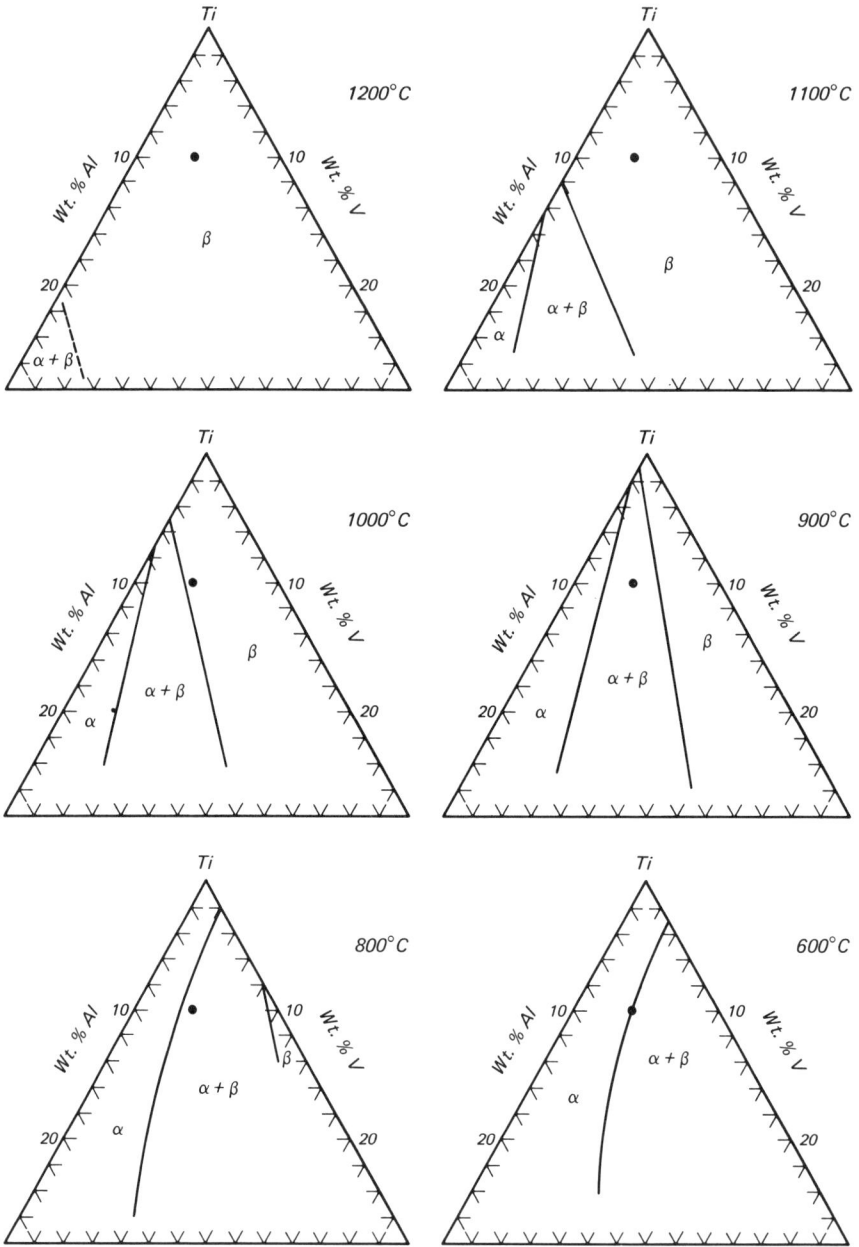

Fig. 3-16. Isothermal (i.e., "horizontal") sections of the Ti-Al-V versus T equilibrium-phase prism at the temperatures indicated [RAU56]. The solid circle (●) represents the alloy Ti-6Al-4V.

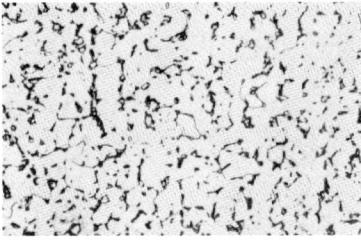

(a) Equiaxed α and a small amount of intergranular β.

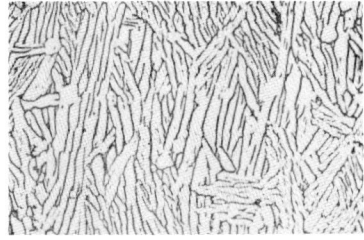

(d) Small amount of equiaxed α in an acicular α (transformed β) matrix.

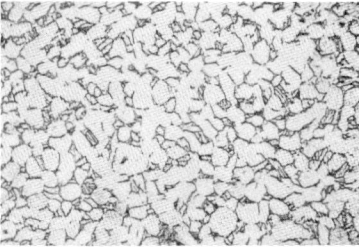

(b) Equiaxed and acicular α and a small amount of intergranular β.

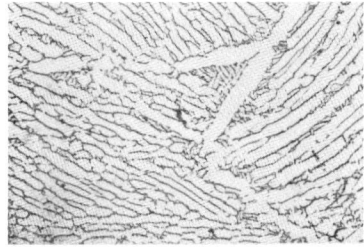

(e) Plate-like acicular α (transformed β); α at prior β grain boundaries.

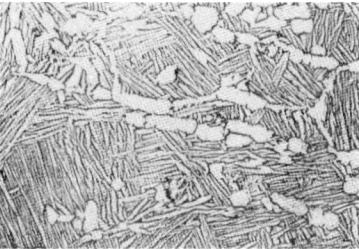

|←—0.1 mm —→|

(c) Equiaxed α in an acicular α (transformed β) matrix.

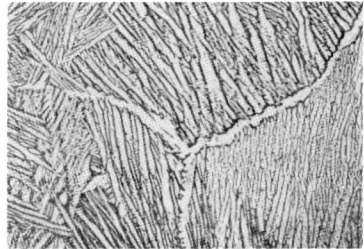

|←—0.1 mm —→|

(f) Blocky and plate-like acicular α (transformed β); α at prior β grain boundaries.

Fig. 3-17. Optical microstructures of Ti-6Al-4V in six representative metallurgical conditions [Woo72, pp. 1-4:72-4, 72-5].

(c) The Near-α Alloy Ti-6Al-2Sn-4Zr-2Mo. The alloy Ti-6Al-2Sn-4Zr-2Mo (Ti-6242) was developed in order to extend the previously existing upper-temperature limit of Ti-alloy service (see *Fig. 1-2*). For this reason, like Ti-8Al-1Mo-1V, it is also describable as a "super-α" alloy. The development philosophy seems to have been to replace the "4V" of Ti-6Al-4V with "2Mo" (the latter being a stronger β stabilizer than V) and to insert

Sn and Zr, which are neutral in this regard. Sn is well known as a solution strengthener; Zr contributed little strengthening but may have been included to improve the stability of the β phase (see subsection (d) below, also Sect. 5.4.1) and/or to assist with homogenization ([FEE70], see Sect. 3.3.4(b)). As with the alloy described in the previous subsection, Ti-6242 functions as a near-α $\alpha + \beta$ alloy.

Further improvement in the elevated-temperature creep strength of Ti-6242 has been obtained by the inclusion in it of 0.2 wt.% Si, which in the form of a silicide precipitate provides dispersion hardening. The alloy so fortified, designated Ti-6242-Si, has been employed as the test alloy in numerous forging research programs [CHE80].

Although the microstructures assumed by this alloy are characteristic of its class, and are generally quite similar to those exhibited by its ternary prototype, Ti-6Al-4V (see *Fig. 3-17*), a wide range of variants can be generated by adjusting the thermochemical-process parameters. In this manner a set of eight microstructural types have been generated by CHEN and COYNE as part of an investigation of the influence of microstructural variation on the mechanical properties [CHE80]. Since the mechanical properties, particularly at elevated temperatures, are very sensitive to the microstructure, a great deal of attention has been given recently to the study of their interrelationships [CHE80]. The two basic microstructures, as depicted in *Fig. 3-18*, are interpretable with the aid of *Fig. 3-19*, a pseudobinary phase diagram with Mo as variable. The $\beta/(\alpha+\beta)$ transus is at about 990°C [WOO72, p. 1-6:72-2]. On cooling from above this temperature, the transformed-β structure varies from martensitic to Widmanstätten $\alpha + \beta$ as the cooling rate decreases from water-quench speeds to those characteristic of air cooling. The structure preserved after an anneal below the $\beta/(\alpha+\beta)$ transus consists of globular α (the original α phase*) plus a transformed version (usually Widmanstätten) of the original β. *Fig. 3-20* shows a set of three typical microstructures selected to represent the effects of air-cooling from anneals at temperatures of 900°C and 980°C. The highest annealing temperature yields the largest volume-fraction of β (see *Fig. 3-19*), which, being the most unstable at ordinary temperatures, yields the finest transformed structure.

A knowledge of the compositions of the α needles and the β matrix of the Widmanstätten β-transformed structure (*Fig. 3-18(a)*), as well as those of any incoherent precipitates which may be present within them, is necessary for a proper interpretation of the mechanical properties. X-ray fluorescence analyses of the latter have been undertaken, and the results

---

*Also referred to as "primary-α".

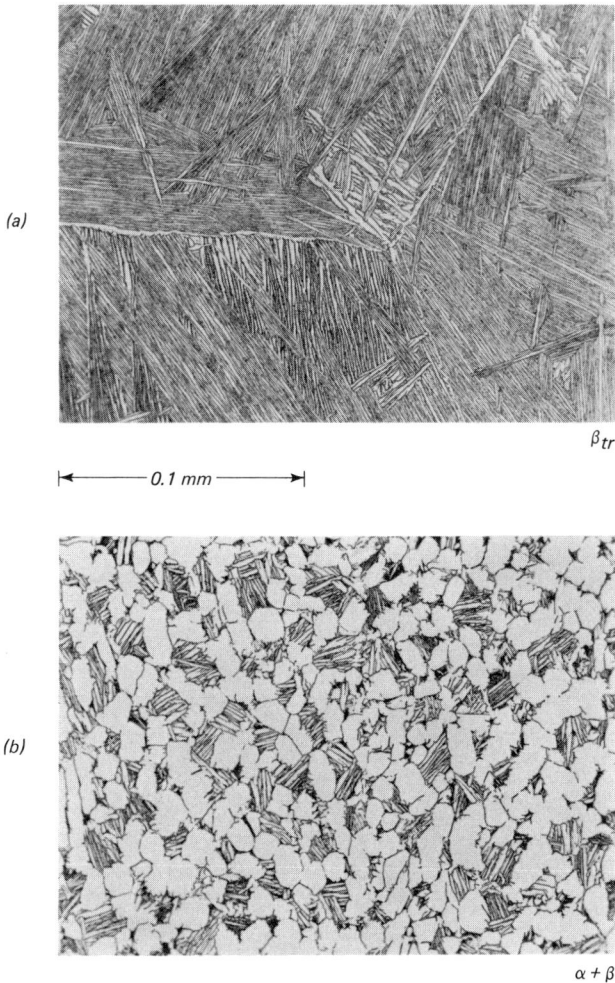

Fig. 3-18. Optical micrographs of Ti-6Al-2Sn-4Zr-2Mo-0.1Si (Ti-6242-Si) in two characteristic metallurgical conditions: *(a)* transformed β (β_{tr} or Widmanstätten $\alpha + \beta$, $(\alpha + \beta)_W$), as a result of the heat treatment 2h/1024°C/air cool; *(b)* $\alpha + \beta$, as a result of 2h/968°C/air cool. Micrographs courtesy of S.L. Semiatin (Battelle).

discussed by CHEN and COYNE [CHE80], while the compositions of the α and β components of the Widmanstätten structure itself have been measured by GEGEL and colleagues, whose results are presented in *Fig. 3-21* and *Table 3-2*. Of particular interest are: (i) the compositional uniformity across the sample exhibited by the "neutral" elements Zr and Sn; (ii) the anticipated preferences of Al for the α needles and Mo for the β matrix;

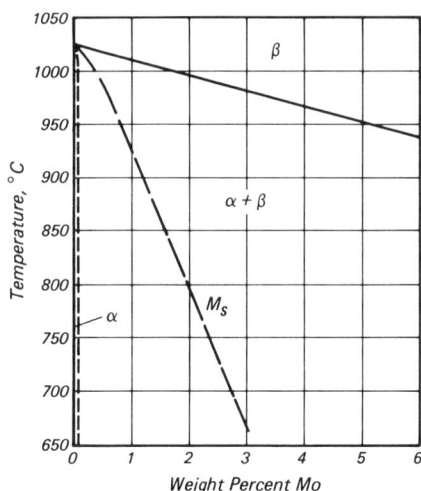

Fig. 3-19. Pseudobinary equilibrium phase diagram for (Ti-6Al-2Sn-4Zr)-XMo for values of X (the wt.% of Mo) between 0 and 6. Source: S.L. Semiatin *et al.*, unpublished research.

Table 3-2. Compositions of the Component Phases in Widmanstätten $\alpha + \beta$-Phase Ti-6242 [Geg80ª]

Component	Ti		Al		Sn		Zr		Mo	
					Composition in wt.% (at.%)					
Average*	86	(85)	6	(11)	2	(1)	4	(2)	2	(1)
β platelet**	78.5	(87)	0.5	(1)	2.0	(1)	4.0	(2)	15.0	(8)
α platelet**	88.5	(88)	5.0	(8)	2.0	(1)	4.0	(2)	0.5	(<1)

*Nominal composition.
**STEM/EDAX analysis.

and (iii) the existence of an "interface phase" characterized by steep compositional gradients of the Al and the Mo.

(d) The Near-β Alloy Ti-6Al-2Sn-4Zr-6Mo. The alloy Ti-6Al-2Sn-4Zr-6Mo (i.e., Ti-6246) is formed from the earlier Ti-6242 by increasing the Mo content in the interests of improved age-hardenability. The additional Mo increases the volume-fraction of the β component at a given temperature, and the presence in it of Sn and Zr, which according to *Fig. 3-21* do not partition between the α and β phases, retards its transformation during cooling, thereby contributing to the hardenability. Detailed descriptions of metallurgy and properties are to be found in [Woo72, pp. 1-7:72-1 *et seq.*].

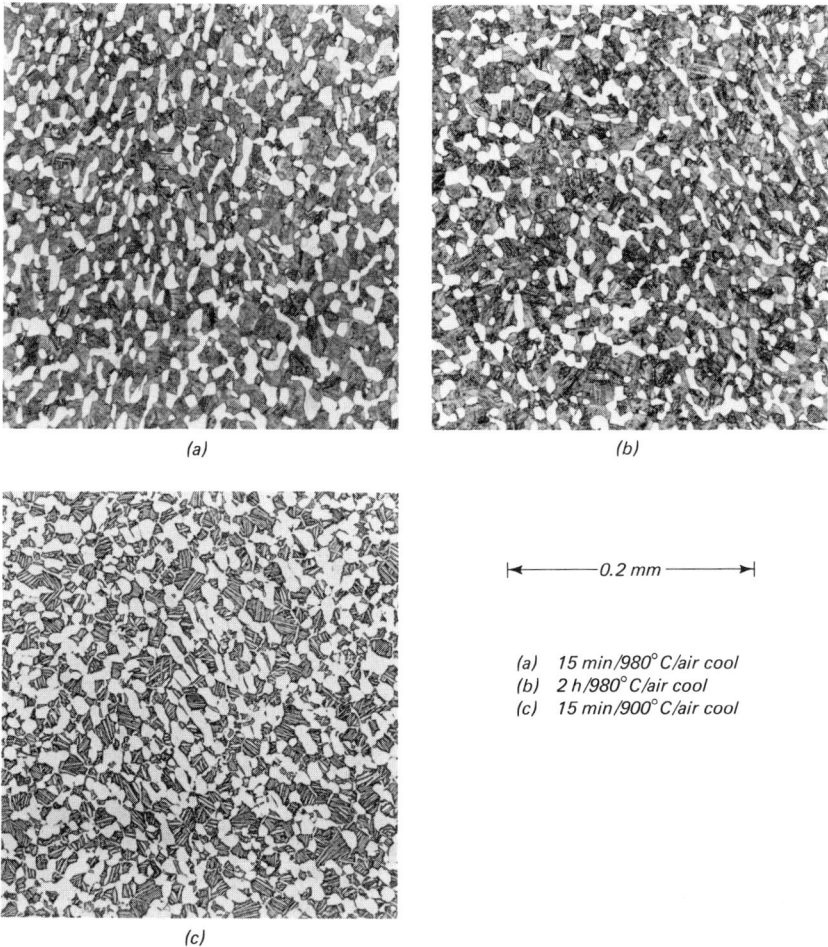

(a)

(b)

|←——————0.2 mm——————→|

(a) 15 min/980°C/air cool
(b) 2 h/980°C/air cool
(c) 15 min/900°C/air cool

(c)

Fig. 3-20. Microstructures in $\alpha + \beta$-phase Ti-6Al-2Sn-4Zr-2Mo-0.1Si (see *Fig. 3-18(b)*) after further exposure to the heat treatments specified. Micrographs courtesy of S.L. Semiatin (Battelle).

3.3.3 The Superconducting Ternary System Ti-Zr-Nb

The Ti-Zr-Nb system is of considerable interest from both practical and pedagogical standpoints in that within it are to be found examples of ω-phase and α-phase precipitation, $\beta' + \beta$ phase separation (near the Ti-Nb edge), and $\beta' + \beta''$ immiscibility (in the Zr-Nb sector of the diagram). A set of equilibrium phase diagrams for this system, based on the work of ALEKSEEVSKII *et al.* [ALE67] and DOI *et al.* [DOI66], who had used the

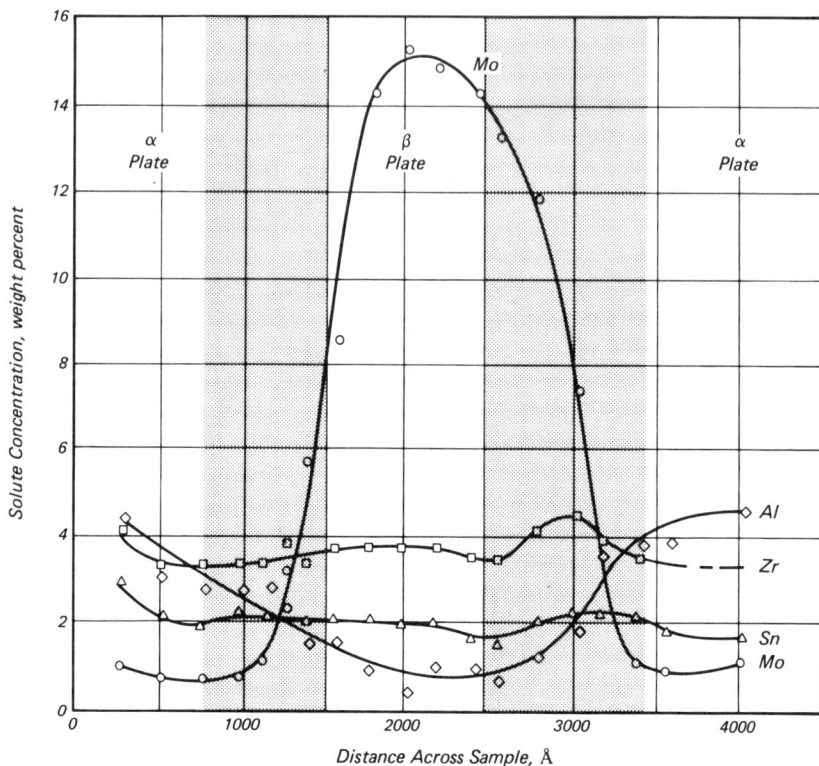

Fig. 3-21. X-ray analysis of the compositional distribution across the $(\alpha+\beta)_W$ structure in Ti-6Al-2Sn-4Zr-2Mo-0.1Si (see *Fig. 3-18(a)*). Source: H.L. Gegel (AFML), unpublished research.

standard techniques of optical metallography and x-ray diffraction, is presented in *Fig. 3-22*. Rather good agreement between the results of these two research groups was obtained, as evidenced by the overall similarity of their diagrams* and, in particular, by the fact that the diagram for 570°C due to Doi *et al.* fits logically between those for 550°C and 600°C obtained by ALEKSEEVSKII *et al.* At 1050°C the ternary systems show unlimited mutual β solid solubility. Below 975°C a region of two-β-phase immiscibility, $\beta'+\beta''$, begins to develop from the Zr-Nb edge, and expands as the temperature falls. Between 700°C and 600°C an $\alpha+\beta$ region develops

---

*In analyzing the x-ray data, Doi *et al.* [Doi66] chose not to distinguish between the two β phases when they coexisted with the α phase, and labelled as $\alpha+\beta$ those regions which ALEKSEEVSKII *et al.* [ALE67] designated (using present terminology) $\alpha+\beta'+\beta''$. In *Fig. 3-22* some reasonable liberties have been taken with the data of Doi for 570°C, after which the diagram falls nicely into position with respect to the surrounding Soviet results.

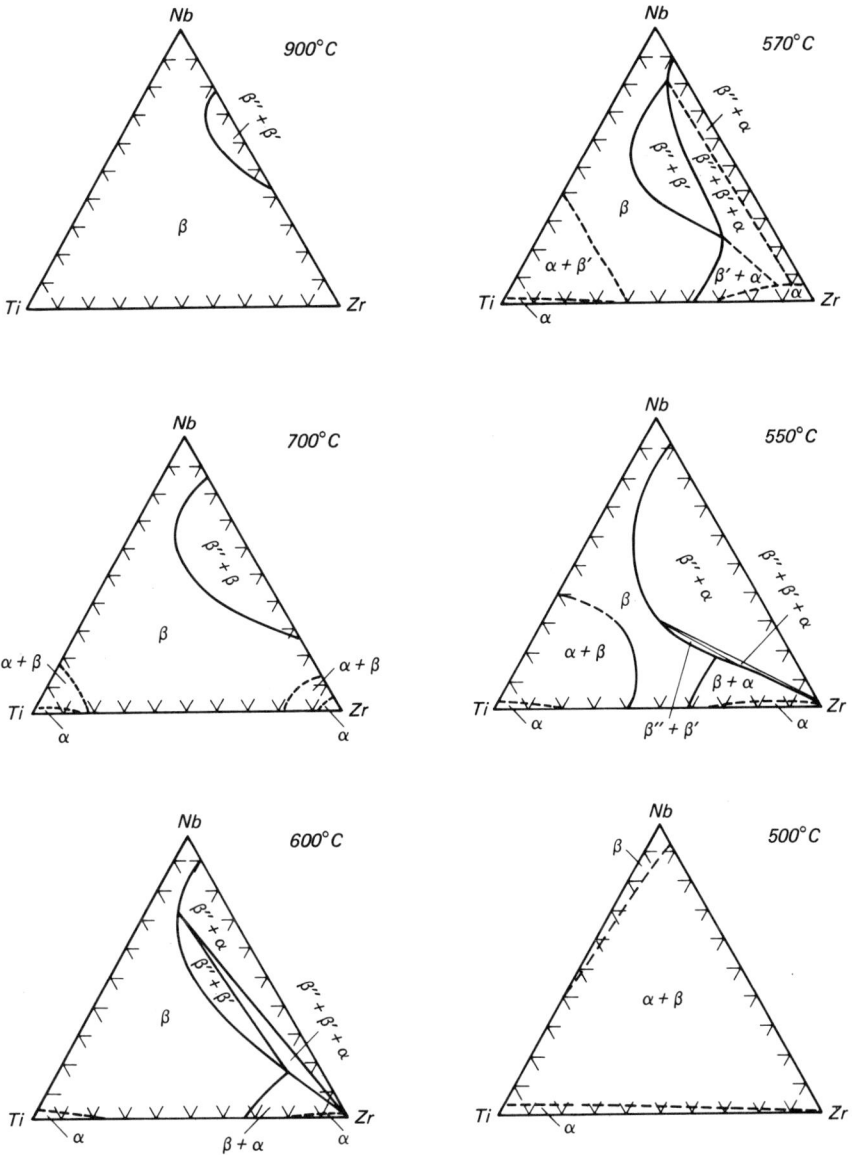

Fig. 3-22. Ti-Zr-Nb equilibrium phase diagram (at.% linear), for temperatures between 900°C and 500°C, based on the work of Doi *et al.* [Doi66] (900°C, 700°C, and 570°C) and ALEKSEEVSKII *et al.* [ALE67] (600°C, 550°C, and 500°C), the latter two diagrams having been modified to take into consideration the absence of appreciable α-phase precipitation from moderate-temperature-aged Ti-Nb alloys with Nb concentration greater than ~40 at.% (see *Fig. 3-10*).

along the Zr-Nb edge separated from the $\beta' + \beta''$ lobe first by a line and, at lower temperatures, by a region $\alpha + \beta' + \beta''$. By this time an $\alpha + \beta$ region is developing from the Ti corner, and another from the Zr corner, severely restricting the remaining region of bcc stability. These changes in phase relationships take place very rapidly with temperature between 570°C and 500°C, at which point an $\alpha + \beta$ phase occupies almost the entire diagram.

3.3.4 β Alloys

The β alloys are characterized by sufficient β stabilizers (transition elements) to ensure the retention of the bcc phase on rapid cooling to room temperature. Strengthening by α-phase particles is then achieved during aging heat treatments. Advantages of β alloys are: (i) cold workability in contrast to the limited room-temperature ductility of the α and $\alpha + \beta$ alloys; and (ii) deep hardenability—i.e., the ability of thick parts formed from them to be hardened [FRO73][PET73]. The first commercial heat-treatable β-Ti alloy was Ti-13V-11Cr-3Al. Subsequently, in order to avoid the difficulties that arose on account of its Cr content, a Mo-containing replacement (β-III) was developed.

(a) The β Alloy Ti-13V-11Cr-3Al. The alloy Ti-13V-11Cr-3Al was for many years the only commercially important β alloy. As a consequence of the very high β-stabilizing content, the β phase is completely retained during even slow cooling ($\sim\frac{1}{2}$ h, for β-annealed material) [Woo72, p. 1-9:72-1]. A pseudobinary equilibrium phase diagram (variable Cr) is given in Fig. 3-23. Hardening to high strength levels is achieved by an aging heat treatment designed to precipitate α phase plus TiCr$_2$. The TiCr$_2$ precipitate derives of course from eutectoidal decomposition associated with the Ti-Cr alloy component, cf. Fig. 3-12; its occurrence in this alloy is responsible for a relatively low aged toughness [PET73] and embrittlement during service at temperatures above 300°C [FEE70]. These drawbacks were to lead to the partial abandonment of "13-11-3" in favor of its Cr-free counterpart, to be described below.

(b) The β Alloy Ti-11.5Mo-6Zr-4.5Sn: "β-III". The alloy Ti-11.5Mo-6Zr-4.5Sn (β-III) contains none of the β-eutectoid formers Cr, Ni, or Cu and consequently does not suffer from the intermetallic-compound embrittlement experienced by its predecessor. β-III was developed in the late 1950s by the then Crucible Steel Company, who also devised melting practices to overcome homogeneity difficulties previously associated with the high melting point and density of Mo [PET71]. β-III is essentially a strengthened Ti-Mo, with which it shares many fundamental metallurgical properties. Solution strengthening of both the α and β phases, when they

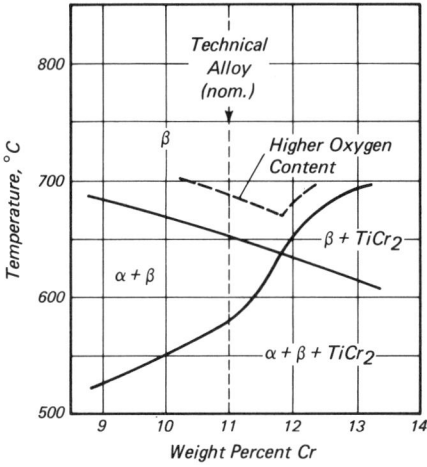

Fig. 3-23. Pseudobinary equilibrium phase diagram for (Ti-13V-3Al)-XCr for values of X (the wt.% Cr) between about 9 and 14 [Woo72; p. 1-9:72-2].

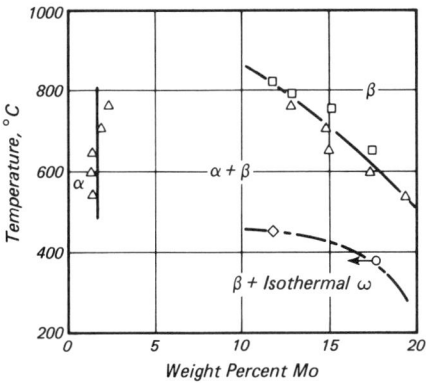

Fig. 3-24. Pseudobinary phase diagram for (Ti-6Zr-4.2Sn)-XMo for values of X (the wt.% Mo) of up to about 19. *Data sources*: (△) partitioning data; (○) behavior of "17.3Mo-β-III" upon 700°F aging; (◇) upper limit of isothermal ω in regular β-III; (□) β transus for 0.28%-oxygen β-III-type alloys [Fro73].

are deliberately produced in it by heat treatment, is conferred by the additions of Sn and Zr (see *Fig. 3-24*). Zr also, apparently, helps to eliminate the inhomogeneity otherwise encountered when melting together elements which differ in density as much as Mo and Ti [Fee70]. A useful review of Ti-Mo-base β alloys has been presented by Ohtani *et al.* of Kobe Steel Ltd., who have also patented a melting technique to overcome the homogeneity difficulties previously encountered in the melting of Ti-Mo alloys [Oht73]. The metallurgical behavior exhibited by β-III is similar to that of binary Ti-12Mo (~6 wt.%) [Boy74]. Thus, decomposition of the bcc phase can take place by both displacive transformation (athermal ω phase) and diffusional transformation (isothermal ω phase and α phase) [Vig82]. Quenching from above the $\beta/(\alpha+\beta)$ transus (755°C) yields a bcc matrix with a fine dispersion of hexagonal ω-phase precipitates no larger than 25 Å [Fee 70][Rac 70]. Binary Ti-Mo alloys with more than 11 wt.% Mo (~6 at.%) exhibit transformation-aided ductility and are capable of more than 30% elongation at room temperature. For similar reasons β-III is also ductile at ordinary temperatures, its cold rollability for example being at least as good as that of commercial unalloyed Ti [Woo 72, p. 1-10: 72-2]. The deformation of β-III within the temperature range 77 K (liquid nitrogen) to 150°C is assisted by twinning and to a larger extent by stress-induced martensitic transformation (to orthorhombic α'') [Fee 70]. In β-III, which is also a deep hardenable alloy, a fine α-phase dispersion is sought. The aging of β-III in order to avoid ω-phase embrittlement in service and to achieve α-dispersion hardening is to be discussed in Sect. 7.14.2.

Nonequilibrium Phases

4.1 General Description

4.1.1 Introduction: Formation of Nonequilibrium Phases

Equilibrium phase diagrams of the type discussed in the previous chapter are usually developed by *deducing the initial states* of alloys which have been quenched to room temperature. The nonequilibrium phases to be considered herein represent the *final* states of such quenching processes. The preceding chapter could scarcely have been written in its present form without anticipating some of the results to be discussed below. In considering the near-α $\alpha+\beta$ alloys, for example (Sect. 3.3.2), it was necessary to point out that quenching from the β field rapidly through the equilibrium $\alpha+\beta$ region resulted in a martensitic structure, while less rapid cooling from the same initial temperature provided an opportunity for α-phase nucleation-and-growth to take place, giving rise to the characteristic Widmanstätten structure, *Fig. 3-18(a)*. The occurrence of these structures has been described in detail by WILLIAMS [WIL73, pp. 1435 *et seq.* and pp. 1460 *et seq.*].

The structure of α-stabilized alloys quenched from the β field is martensitic. When quenched from below the $(\alpha+\beta)/\alpha$ transus the structures found are of course simply the frozen-in untransformed results of equilibration at the pre-quenched temperature.

The structures assumed by rapidly β-quenched binary Ti-TM alloys are mapped in *Fig. 4-1*. Below a start temperature, M_s, the bcc structure begins a spontaneous allotropic transformation by means of a complicated shearing process to a structure known as martensite and designated α' or α'' depending upon whether the transformation product is hcp or orthorhombic. When the distinction between α' and α'' is unimportant, the martensites are to be herein represented collectively by the notation α^m. Being of second order, the martensitic transformation is anticipated by a regime of structural fluctuations called diffuse ω phase. As represented in *Fig. 4-1*, the ω phase, as a result of very rapid quenching, exists as a crystalline precipitate plus a fluctuating component within a narrow composition range overlapping the boundary of the martensite phase. In practice,

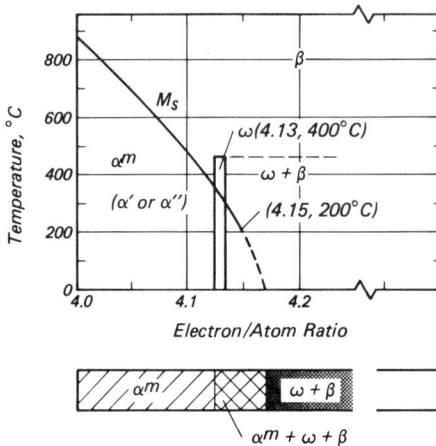

Fig. 4-1. Schematic representation of the occurrences of the martensitic phases α' and α'' (i.e., α^m, collectively) and the ω phase in Ti-TM alloys. Both "quenched data" and "aged data" are included, see also *Table 4-1* (α^m phases) and *Table 4-2(a)* (ω phase).

however, the range over which it occurs during the brine quenching of macroscopic samples is quite broad and is depicted in *Fig. 4-1* as a region of gradually diminishing precipitate abundance. The free energy of α^m is lower than that of ω; consequently, during the partial martensitic transformation of an alloy in which ω phase is also able to form, the martensite needles generally consume any ω-phase precipitates which lie in their paths.

4.1.2 The Quenching Process

In studies of quenched microstructures, an important but not always accomplished goal is the control and quantification of the quench rate. If the quench is too slow, diffusional processes intervene to obscure the result. When the primary aim is to study microstructure (rather than the production of material for physical- or mechanical-property testing) and the fastest possible quench rates are mandatory, thin foils are generally heated in a controlled environment and subjected to *in situ** gas or liquid quenching. For example in HICKMAN's method, rolled strips self-heated under high vacuum to 1250°C by the passage of direct current were quenched by admitting He gas to a pressure of 0.1 atm, restoring sample temperature, and then switching off the current [HIC68, HIC69a]. Such techniques are generally capable of quench rates of 50°C s^{-1} to 2×10^4°C s^{-1}, and in HICKMAN's case about 10^3°C s^{-1} was claimed. Helium is about three times as effective a quench medium as Ar under the same conditions.

---

*In which the quenching medium is introduced into the furnace space containing a fixed sample.

BROWN, JEPSON, and HEAVENS [BRO65] heated indirectly under vacuum specimens varying in thickness from 0.002 to 0.20 in; they then applied a 150-700 torr head-pressure of Ar to suppress the boiling of the iced water or refrigerated calcium chloride solution subsequently admitted to quench the sample. In this way quench rates of 2.5×10^4 to 2×10^5 °C s^{-1} were achieved [JEP70]. BALCERZAK and SASS [BAL72], whose results are discussed below, attached rolled specimens (0.002-0.003 in thick) to an Inconel specimen holder by means of which they could be transferred from the hot zone of a vacuum resistance furnace to a waiting pool of water-cooled silicone oil. Although they were known to be rapid, the quench rates achieved by this method were not specified. The quench rates achievable by all of these methods are of course much faster than those obtained during the ice-brine quenching of the massive samples (up to 40 g) needed for mechanical- and physical-property study (especially low-temperature specific heat). Accordingly, some discrepancies must be expected between the microstructural results obtained from thin foils as compared to those derived from the quenching of bulk specimens.

The measured M_s temperature for a given alloy composition is itself a function of quench rate. In Ti-Nb(5 at.%), for example, JEPSON et al. [JEP70] noted that the M_s temperature decreased from 760°C to 710°C as the cooling rate increased from 10^{-3} to 10°C s^{-1}, but that once a critical cooling rate of 32°C s^{-1} was exceeded, M_s was independent of the cooling rate. The critical threshold itself was a function of alloy composition and decreased from 200°C s^{-1} to ~0.4°C s^{-1} as the Nb content increased from 0 to 15 at.%.

In Ti-TM alloys, as with other systems, the quench rates necessary to achieve structural transformation while preserving compositional homogeneity are strongly constitution dependent. Thus, whereas bulk dilute alloys of Ti with early transition elements can be water quenched without evidencing serious decomposition, the same is not true of alloys such as Ti-Fe, -Ni, and -Co, whose anomalous superconducting properties (to be discussed below) could be partially interpreted in terms of compositional, hence structural, segregation. The pronounced differences between the properties of the quenched dilute Ti-V, -Nb, etc. alloys and those of Ti-Fe, -Co, and -Ni can be simply explained in terms of differences among the solute tracer diffusion coefficients in β-Ti at 1000°C. As shown in *Fig. 2-18*, the diffusion coefficients of V, Nb, and Mo are less than 1.3×10^{-9} cm^2 s^{-1}, while those of the Fe-group elements are 60×10^{-9} cm^2 s^{-1}. The extreme examples are Co on one hand and Mo on the other; their diffusion coefficients (β-Ti, 1000°C) are in the ratio 200:1. This must be taken into consideration when comparing the properties of the two classes of Ti-base alloy, and in selecting a quenching technique.

4.1.3 Stability Limit of the β Phase in Titanium-Transition-Metal Alloys

The optical microstructures and composition ranges of the martensites characteristic of ice-brine-quenched massive samples of Ti-V, -Nb, -Mo, and -Fe are exemplified by *Fig. 4-2*. Compositionally, Ti-Fe and Ti-Nb are extreme examples, their M_s curves bounding those for all other measured Ti-TM alloys and intersecting a 200°C isothermal, for example, at 3.3 and 20.5 at.%, respectively, [Zw174, p. 174].

In *Table 4-1* are listed the $M_{s,200°C}$ compositions of eight Ti-TM alloys bounded constitutionally by Ti-Fe and Ti-Nb, together with the corresponding conventional electron/atom ratios. Quite remarkable is the fact that all the e/a ratios, except for those of Ti-Co and Ti-Ni (which may be exceptional cases), lie within ±0.03 of a common value, 4.15, suggesting that the martensitic transformation in Ti-TM alloys is of common origin and related to electronic factors. Electronic and lattice properties can, at least in principle, be coupled through calculations of the elastic constants.

Table 4-1. Compositions of the $M_{s,200°C}$ Intercepts Expressed in Terms of Conventional Electron/Atom Ratio

Solute group number, GN	Solute element	Concentration corresponding to M_s at 200°C, at.%* c	Conventional** e/a based on group number	
V	V	13.3	4.13	
	Nb	20.5	4.21	
	Ta	19.1	4.19	
VI	Cr	6.0	4.12	
	Mo	6.7	4.13	
	W	8.2	4.16	
VII	Mn	5.0	4.15	
VIII	Fe	3.3	4.13	
	Co	6.0	4.24	4.30†
	Ni	7.6	4.30	4.46†

Mean†† 4.15 ± 0.03

*After Zwicker [Zw174, p. 174].

**Calculated according to: $e/a = 4 + \Delta GN \dfrac{c}{100}$, where $\Delta GN = GN_{solute} - GN_{Ti}$,

$$= GN_{solute} - 4.$$

†Based on number of valence $(s+d)$ electrons.
††Excluding Co and Ni.

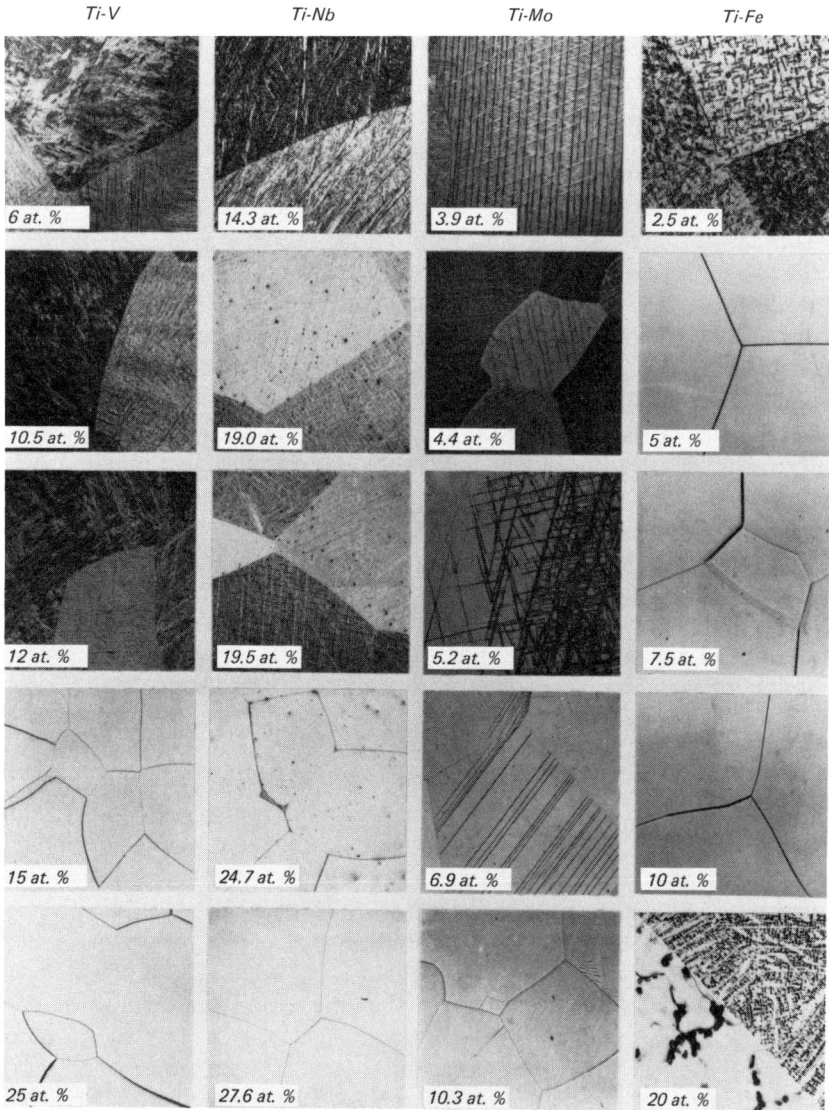

Fig. 4-2. Optical micrographs from massive tokens of Ti-TM alloys, quenched from the β phase into iced brine, showing the transitions from the α^m to the $\omega + \beta$ regimes with increase in solute concentration (or e/a ratio). Compositions of Ti-V and Ti-Fe are nominal, those of Ti-Nb and Ti-Mo are analyzed. Magnifications of the original 9×9 cm^2 micrographs were 50×; a 300× micrograph of Ti-Fe(20 at %) is also shown [CoL84].

A connection between the end of bcc stability and the vanishing of the elastic shear modulus, $C' = (C_{11} - C_{12})/2$, has already been considered in Sect. 3.1.4 (see *Fig. 3-3*) and is considered again in Sect. 5.4.3 and in Sect. 6.7 with particular reference to the Ti-Cr and Ti-Mo systems [COL72, COL72[a], COL73].

In discussing the occurrence of ω phase in Ti alloys, *Fig. 4-1* makes a useful starting point. As represented therein, a narrow region exists in which ω phase appears *athermally* during very rapid quenching from the β phase. Over a broader composition range ω phase will occur as a precipitation product of β decomposition during moderate temperature ($\gtrsim 400°C$) *isothermal* aging; alloys *rapidly quenched* into this broader region are host to a "diffuse ω phase", so-called because of the existence of straight or curvilinear lines of diffuse intensity in selected-area electron diffractograms [LUH70][BAL71][SAS72]. The occurrence, composition, and structure of ω phase, and its relationship to the competing α and β phases, have been reviewed by HICKMAN [HIC69] and SASS [SAS72], who made reference to important earlier work, including the often-quoted studies of SILCOCK [SIL58] and BAGARIATSKII *et al.* [BAG59].

During moderate-temperature ($\gtrsim 400°C$) aging for several days a metastable equilibrium $\omega + \beta$ state is attained, analysis of which yields the compositions of the ω (solute-lean) and β (solute-rich) phase boundaries. The ω-phase regime does not extend to zero solute concentration, but is tightly confined by the nearby lower-energy martensitic-phase field. As a result of the limited composition range of the athermal ω phase, good agreement is obtained between the "saturation composition of ω phase after aging at 400°C" [HIC69[a]] and the "solute concentrations which yield ω phase on quenching" [BAG59], as shown in *Table 4-2*. The latter is of course a difficult quantity to determine accurately, since away from the narrow zone of athermal ω phase, rapidly quenched alloys support the diffuse ω phase, and more slowly quenched larger samples will contain isothermal ω. *Table 4-2* gives the conventionally calculated e/a ratios for ω phase, separated into two listings according to the source and the method of data acquisition. It is remarkable to note that in each case the critical e/a ratios for the eight alloys listed are constant (within 20% in terms of $\Delta(e/a)$), and that the two independent mean values (which differ by <0.01) agree within experimental scatter. Secondly, a comparison of *Table 4-1* and *Table 4-2*, which yield, respectively, $\langle e/a \rangle_{\alpha^m} = 4.15 \pm 0.03$ and (from an overall mean) $\langle e/a \rangle_\omega = 4.13 \pm 0.03$, emphasizes that athermal ω phase occurs at the threshold of martensitic transformation and suggests that the ω and α^m transformations are interrelated through a common electronic mechanism. The results also reconfirm the validity of *Fig. 4-1*.

The results of a literature study of collected transformation data for

Table 4-2(a). Compositions of the $\omega/(\omega+\beta)$ Data Points Expressed in Terms of Conventional Electron/Atom Ratio

Solute group number	Solute element	Saturation composition of ω phase (at.%) after aging at ~400°C [Hic69a]	Solute concentration (at.%) in Ti alloys for which ω phase is formed on quenching [Bag59]	Conventional e/a based on group number (see Table 4-1)	
				"Aged data"	"Quenched data"
V	V	13.8 ± 0.3	[— 13	4.14	[— 4.13
	Nb	~9 ± 2	— 18]	4.09	— 4.18]
VI	Cr	6.5 ± 0.2	[— 7	4.13	[— 4.14
	Mo	4.3 ± 0.4	— 4.5	4.09	— 4.09
	W	—	7.5]	—	4.15]
VII	Mn	5.1 ± 0.2	[— 5.5	4.15	[— 4.16
	Re	—	4.5]	—	4.14]
VIII	Fe	4.3 ± 0.2	[— 3]	4.17	[— 4.12]
				Mean 4.13 ± 0.03	Mean 4.14 ± 0.03

Table 4-2(b). Compositional Limits of the $\omega + \beta$ Phase in Quenched Alloys of Ti with Other Transition Elements [GUS82]

Alloy		Solute concentration range, at.%		PAULING e/a-ratio range	
Solvent	Solute	Minimum	Maximum	Minimum	Maximum
Ti	V	14	21	4.14	4.21
Ti	Nb	12	20	4.12	4.20
Ti	Cr	—	11	—	4.20
Ti	Mo	6	10	4.11	4.18
Ti	Fe	4	10	4.07	4.18
Ti	Ru	5	10	4.09	4.18
Ti	Ni	7	—	4.12	—
Ti-V(2.5 at.%)	Ru	5	—	4.13	—
Ti-V(5 at.%)	Ru	3.5	7.5	4.13	4.20
Ti-V(7.5 at.%)	Ru	2.5	6	4.13	4.20
Ti-V(10 at.%)	Ru	1.5	4	4.13	4.18
Ti-V(15 at.%)	Ru	—	2.5	—	4.20

nine Ti-TM alloys, conducted by LUKE, TAGGART, and POLONIS [LUK64], independently of that of BAGARIATSKII *et al.* [BAG59] and prior to that of HICKMAN [HIC69[a]], and indexed in terms of an e/a ratio based on PAULING valence [PAU56] rather than group number, led to a similar conclusion. In conducting this survey of the limits of bcc stability, LUKE *et al.* [LUK64] chose not to distinguish between martensite and ω phase as the transformation product. But after reviewing the original references cited therein and separating the results into "α^m-limited" and "ω-phase-limited" β phase as in *Table 4-3*, one is led again to the conclusion—supported by the data of *Table 4-1* and *Table 4-2* considered jointly—that: (i) in terms of e/a, the threshold for the α^m transformation (presumably $M_s \cong$ room temperature) is constant to within about ± 0.03; (ii) the composition of athermal ω phase expressed as an electron/atom ratio is also constant to within about ± 0.03; and (iii) the martensitic and ω-phase "thresholds" agree to well within that scatter.

The significance of e/a as an index of the limit of bcc stability in Ti-TM alloys was emphasized by LUKE, TAGGART, and POLONIS [LUK64], who were probably the first to point explicitly to the importance of electronic factors in the stability of bcc Ti-TM alloys.

The presence of athermal ω-phase particles, too small ($\sim$20-40 Å) to be detected optically, can be unequivocally diagnosed using TEM and selected-area electron diffractometry (SAD). By way of example, their occurrence in ice-brine-quenched Ti-Mo(5 at.%), which is just on the edge of the quenched martensitic regime (*Fig. 4-2*), has been confirmed using these techniques, the results of which are shown in *Fig. 4-3*.

Table 4-3. Solute Concentrations and Electron/Atom Ratios for 100% β Stabilization in Ti-TM Alloys—After Luke et al. [Luk64]

Solute group number	Solute element	Critical solute concentration, at.%	PAULING valence [Pau56]	Conventional e/a based on group number (see Table 4-1)	PAULING e/a	Phase with respect to which the β phase was considered stable*	References
V	V	15	5	4.15	4.148	ω	[Bro55][Sil58]
	Ta	15-21**	5	4.15	4.148	m	[May53]
VI	Cr	7.5	6	4.15	4.147	ω	[Spa58]
	Mo	7.2	6	4.14	4.143	m	[Del52]
	Mo	7.4	6	4.15	4.148	ω	[Sil58]
	W	7	6	4.14	4.14	m	[Ots61]
VII	Mn	5.5	6	4.17	4.118	ω	[Jaf58]
		7.5	6	4.23	4.148	m or ω	[Fro54]
		8	6	4.24	4.158	ω	[Ler60]
VIII	Fe	6-8†	6	4.28	4.139	ω	[Pol55]
	Co	6	6	4.24	4.12	m	[Swa58]
		7.5	6	4.29	4.148	ω	[Yak61]
		5-7†	6	4.24	4.12	m	[Orr55]
	Ni	7	6	4.28	4.14	m	[Mar53]
		7-8†	6	4.29	4.15	ω	[Bar60]

Mean conventional e/a → 4.20 ± 0.06 or 4.22 ± 0.06

Mean PAULING e/a → 4.14 ± 0.01 or 4.14 ± 0.01

*i.e., whether an M_s composition (m) or absence of ω phase (ω) was considered.

**The lower value is taken.

†Mean taken.

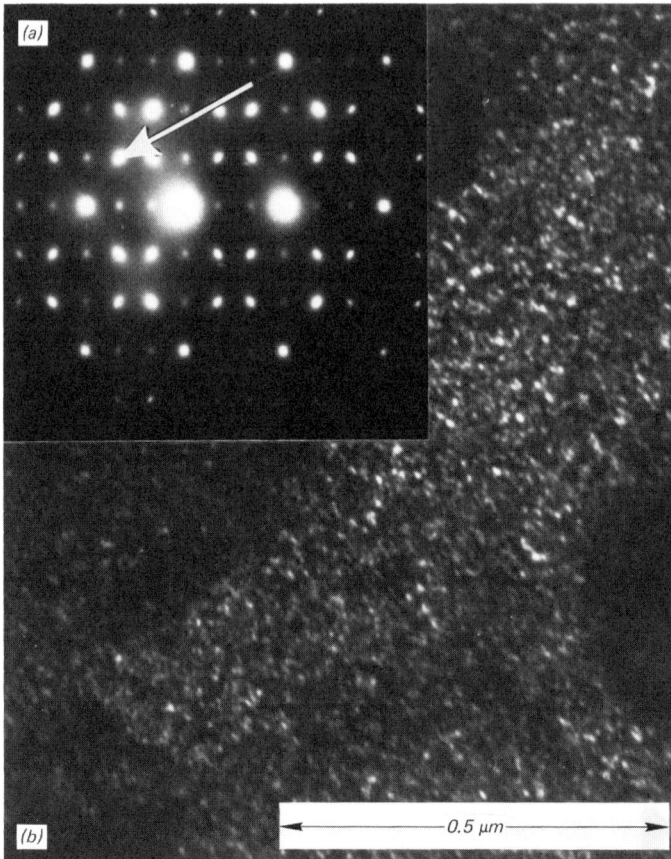

Fig. 4-3. Transmission electron micrographs of Ti-Mo(5 at.%) quenched into ice brine from the β field. *(a)* Electron diffractograph showing a superposition of two principal spot patterns: a rectangular arrangement of round spots originating from the β matrix, and groups of elongated spots originating from the ω phase. *(b)* Dark-field electron micrograph of the ω-phase precipitate originating from the diffraction spot indicated by the arrow in the diffractograph. Using the dark-field technique, the bright patches of the photograph are specific to the ω phase, however, only one-quarter of the precipitate is visualized at one time [Col71^{d}].

The occurrence of athermal ω phase in such close proximity to the martensitic phase boundary is a phenomenon of general validity and of fundamental importance in the theory of the ω transformation (to be considered later in more detail). Although from the standpoint of the crystallographer the absence of a habit-plane description precludes the defining of the ω-phase transformation as martensitic [Wil73], both transformations, according to Suzuki and Wuttig [Suz75] and Clapp

[CLA 73], possess a common ingredient in the form of a soft-phonon instability. Arguments supporting this view are developed in subsequent sections.

On general thermodynamic grounds it can be shown that after quenching through an equilibrium phase boundary, the boundary of a metastable phase is encountered, just outside of which a region of statistical fluctuations is expected. Near a *structural* transformation, *strain* fluctuations would be appropriate. It is known from the neutron diffraction studies of Moss *et al.* [Mos73] and the Mössbauer effect measurements of BATTERMAN *et al.* [BAT73] that ω phase possesses a fluctuating component. It is, therefore, tempting to suggest that the dynamic (or "diffuse") ω is the critical opalescence of the athermal precipitate and/or martensite.

4.2 Formation and Structures of the Martensitic Phases in Titanium Alloys

The word martensite, named for Professor A. Martens, was originally adopted by metallurgists to define the acicular structure in quenched carbon steel that was responsible for its outstanding hardness [COH51]. The occurrence and structures of martensites in numerous other alloy systems have been described by COHEN [COH51] and by BILBY and CHRISTIAN [BIL56]. Detailed discussions of martensitic transformations, but with particular reference to Ti alloys, have been offered by the McQUILLANS [MCQ56, Chap. 9], MARGOLIN and NIELSEN [MAR60], HAMMOND and KELLY [HAM70], and OTTE [OTT70]. Unalloyed Ti transforms martensitically from bcc to hcp during cooling through its $\beta \rightarrow \alpha$ allotropic transformation temperature, 882.5°C. In alloys of Ti, the equilibrium α and β fields are separated by a two-phase, $\alpha + \beta$, region and the $\beta \rightarrow \alpha^m$ transformation temperature,* M_s, is composition dependent. In α-stabilized alloys, typified by Ti-Al, M_s may lie a little below the $(\alpha + \beta)/\alpha$ transus [JEP70]; in the β-stabilized alloys it always lies within the $\alpha + \beta$ field. For some recent information on martensitic transformations in the β-isomorphous systems Ti-V, Ti-Nb, and Ti-Mo, and the manner in which the transformed structures revert to β or decompose on aging, the papers of DAVIS, FLOWER, and WEST [DAV79, DAV79^a][FLO82] should be consulted.

4.2.1 The Martensitic Transformation in Titanium Alloys

In contrast to nucleation-and-growth types of phase change which rely on thermally activated atomic diffusion, martensitic transformations involve a cooperative movement of atoms resulting in a microscopically

---

*The symbol α^m is used herein as a short-hand notation for the product of martensitic transformation, whether it be α' or α''. M_s refers to the start of the transformation during cooling; M_f usually designates its finish.

homogeneous transformation of one crystal lattice into another. The ideal martensitic process itself is not thermally activated and takes place at high temperature-independent speeds; but in practice a clear-cut separation of transformation processes into "nucleation-and-growth" and "martensitic" is generally not possible. Although in pure metals, such as unalloyed Ti, a simple athermal martensitic transformation is conceivable, in an alloy the situation is more complicated. The *long-range effect* of alloying, described for example in terms of changes of the elastic parameters, may simply be to change the conditions under which athermal transformation takes place. The *local effect* of alloying, on the other hand, is to inhibit the movement of atomic planes and thereby: (i) to reduce the distances over which atomic regions can cooperate, which perturbs the microstructure of the transformation product; and (ii) to reduce the speed of the transformation, thus bringing it into competition with the nucleation-and-growth mechanism. The influence of solute concentration on the kinetics of the transformation can be seen in the results of the experiments of JEPSON, BROWN, and GRAY [JEP70] on a series of Ti-Nb(0-17.5 at.%) alloys. In studies of the influence of cooling rate, say $r \equiv dT/dt$, on the transformation temperature, M_s, they were able to show that the critical cooling rate, r_c (at rates faster than which M_s became independent of r), decreased with increasing Nb concentration. Thus, for example, whereas in unalloyed Ti, $r_c = 10^2 °C\,s^{-1}$, the addition of 15 at.% Nb reduced it to $0.3°C\,s^{-1}$. In considering what might be referred to in Ti-TM alloys as a "solute-controlled transformation at solute-inhibited transformation speed", in which athermal shear competes with diffusion, a given quench rate cannot be expected to yield similar transformed structures for all concentrations and all types of Ti-TM alloys. The properties of the quenched product will depend on several variables, one of the most important of which is the diffusion coefficient of the particular solute species in β-Ti. *Fig. 2-19*, which compares the 1000°C-diffusion coefficients of eight transition elements in β-Ti, shows that a quench rate that will completely suppress diffusion in Ti-Nb ($D_{Nb} = 1.3 \times 10^{-9}\,cm^2\,s^{-1}$) and Ti-Mo ($D_{Mo} = 0.3 \times 10^{-9}\,cm^2\,s^{-1}$) may be quite inadequate to do so in Ti-Fe, Ti-Ni, and Ti-Co (for which $\langle D \rangle_{Fe,Ni,Co} = 61.6 \times 10^{-9}\,cm^2\,s^{-1}$). Against this background it is possible to believe in the existence of a series of quenched structures extending from "bulk martensite" through "acicular martensite" to the Widmanstätten structure defined below.

4.2.2 Morphology of the Martensites

When conditions are particularly favorable, transformation from β to α^m takes place completely, on a large scale, and with considerable structural coherence. The result is the so-called "massive martensite" (otherwise known as packet, or lath, martensite), which consists of large irregular zones on the scale of 50-100 μm, subdivided into parallel arrays of fine platelets less than 1 μm across, *Fig. 4-4.* In massive martensite, the lack of retained β phase prevents direct determination of the habit plane. With increasing solute concentration, the coherence between the platelets, which would otherwise make up a massive colony, is lost. The result of this is a partially disordered array of individual platelets referred to as "acicular martensite", *Fig. 4-5.* Further increase in solute concentration prevents a

Fig. 4-4. An example of "massive martensite". *Specimen*: Ti-1.78Cu quenched from 900°C. *(a)* Optical micrograph showing large colonies. *(b)* Electron micrograph showing individual plates within the colonies [Wɪʟ73]. Micrographs courtesy of J.C. Williams (Carnegie-Mellon University).

Fig. 4-5. An example of "acicular martensite". *Specimen*: Ti-12V quenched from 900°C. *(a)* Optical micrograph. *(b)* Electron micrograph showing lenticular-shaped plates, some of which are internally twinned [Wɪʟ73]. Micrographs courtesy of J.C. Williams (Carnegie-Mellon University).

complete transformation from taking place, and β phase trapped between the platelets of the acicular martensite enables direct habit plane determination to be accomplished. Not far removed from the β-plus-acicular-martensite quenched structure is the Widmanstätten arrangement consisting of groups of α-phase needles lying with their long axes parallel to the {110} planes of the parent retained β, *Fig. 4-6*. Widmanstätten $\alpha + \beta$ (or Widmanstätten α, if the focus is primarily upon α-phase precipitation), which is characteristic of dilute Ti-TM alloys or near-α $\alpha + \beta$ alloys (such as Ti-6Al-4V) appropriately cooled, is usually treated as a product of α-phase nucleation and growth. It is introduced into this chapter on non-equilibrium phases since, when solute diffusion coefficients are sufficiently large, either intrinsically so (large frequency factor, D_0) or if the temperature is high, diffusional processes compete with diffusionless martensitic transformation during the quenching of β-Ti alloys [CHR65][JEP70] [FLO82].

The optical microstructures of a series of β-quenched Ti-TM alloys are shown in *Fig. 4-7*. The five alloys Ti-Mo, Ti-V, Ti-Nb, Ti-Mn, and Ti-Fe have comparable e/a ratios, had similar masses, and were ice-brine

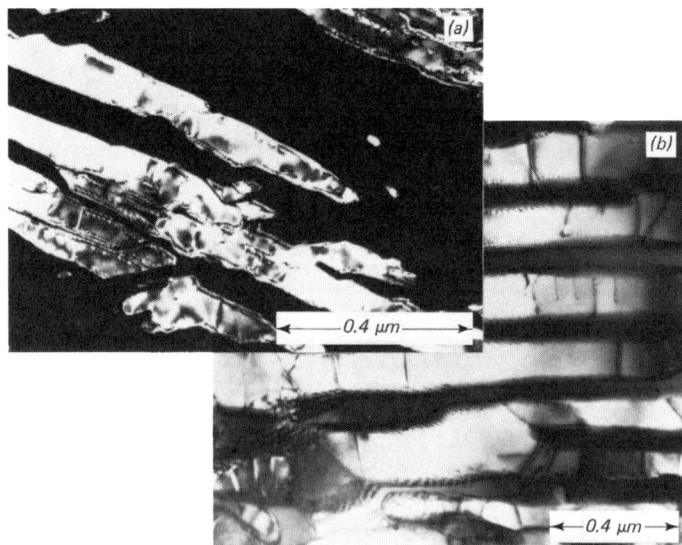

Fig. 4-6. An example of Widmanstätten α phase. This type of α phase is found in near-α $\alpha + \beta$ binary alloys and in alloys with substantial amounts of Al, e.g., Ti-6Al-4V [WIL73]. *(a)* Dark-field micrograph showing absence of internal structure in the platelets. *(b)* Bright-field micrograph showing the α platelets separated by the β matrix. Micrographs courtesy of J.C. Williams (Carnegie-Mellon University).

Ti-Mo (4.5)/an.   4.09   Ti-V (9)/nom.   4.09   Ti-Nb (8.9)/an.   4.09

Ti-Mn (2.7)/an.   4.08   Ti-Fe (2.5)/nom.   4.10

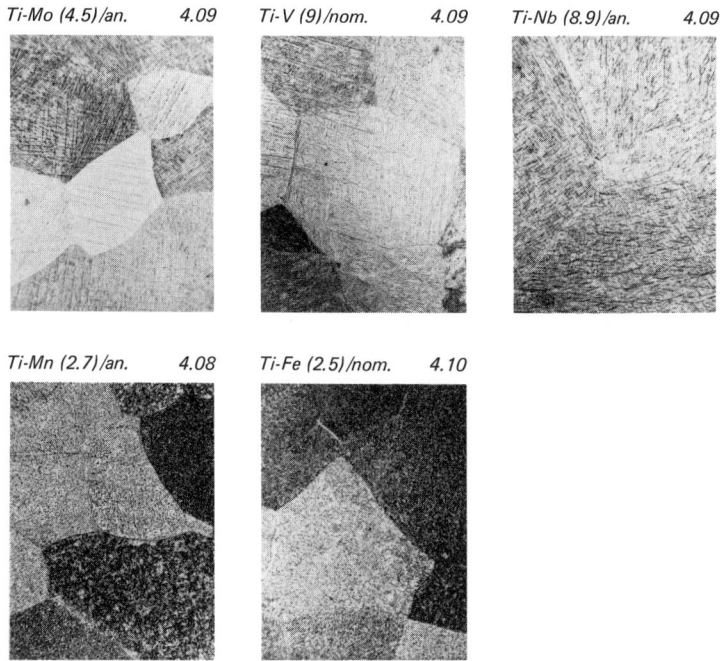

Fig. 4-7. Optical micrographs of five low-concentration Ti-TM alloys after quenching into iced brine from the β phase, arranged in ascending order of solute atomic diffusion coefficient in β-Ti (see *Fig. 2-18*). Indicated are the analyzed solute concentrations (nominal in the cases of Ti-V and Ti-Fe) and the conventional electron/atom ratio ($\cong 4.09$ = const.). Magnifications of the original 11.5×9 cm^2 micrographs, 50×.

quenched under similar conditions. Noticeable in the photomicrographs of these alloys, which are arranged in ascending order of solute diffusion coefficient in β-Ti (*Fig. 2-19*), is what appears to be a gradual transition from a diffusionless to a diffusion-influenced as-quenched structure on proceeding from Ti-Mo to Ti-Fe. These results suggest that atomic diffusion coefficient must be introduced as a scaling factor when estimating the effects of quench rate on martensitically transformable Ti-TM alloys.

Metastable-bcc* Ti-TM alloys will also transform under the application of mechanical stress. Although some confusion has arisen in the past over the structure of the deformation product, the situation has been adequately clarified by WILLIAMS [WIL73]. The β-quenched solute-lean but untransformed Ti-TM alloys are characterized by low shear moduli. In such

---

*Alloys β-quenched into the $\alpha + \beta$ field but without intersecting M_s.

Fig. 4-8. An example of deformation martensite and/or twinning. *Specimen*: Ti-Mo(5 at.%) quenched from the β phase and deformed 23% by compression. Magnification of the original 9×9 cm^2 micrograph, 50×.

an alloy, which may be represented in the phase diagram by a point close to M_s, the application of stress will trigger a transformation to "deformation-induced" or "stressed-induced" martensite. Farther away from the M_s transus the bcc lattice responds directly to the influence of an applied stress by twinning. This, as pointed out by SUZUKI and WUTTIG [SUZ75], can in the present context be regarded as a special case of martensitic transformation to a crystallographically equivalent structure. Indistinguishable optically from deformation martensite, mechanical twins are easily identified by diffractometry — their structure, of course, being identical to that of the parent lattice. Stress-induced martensitic transformation and twinning in Ti-Mo alloys have been recently discussed by OKA and TANIGUCHI [OKA80]. Deformation-induced orthorhombic martensite produced by compressively deforming a β-quenched Ti-Mo(5 at.%) alloy is depicted in *Fig. 4-8.*

4.2.3 Structure of the Martensites

The transformed structure assumed by the pure elements Ti, Zr, and Hf, and by the dilute Ti-TM alloys in general, is hcp and has been assigned the symbol α' [BAG59][WIL73]. Otherwise, quenched Ti-TM alloys with compositions exceeding certain limits, which differ from system to system, transform to an orthorhombic martensite designed by α''. The α'/α'' compositional boundaries for the alloy systems Ti-V, Ti-Nb, Ti-Ta, Ti-Mo, Ti-W, and Ti-Re have been determined by BAGARIATSKII *et al.* [BAG59], whose results are summarized in *Table 4-4*. The results of more

Table 4-4. Compositions and Electron/Atom Ratios of the α'/α'' Boundary in Some Ti-TM Binary Alloys [BAG59]

Element		α'/α'' Boundary		
	wt.%	at.%	ΔGN*	Conventional e/a ratio*
V	9.4	8.9	1	4.09
Nb	10.5	5.7	1	4.06
Ta	26.5	8.7	1	4.09
Mo	4	2.0	2	4.04
W	8	2.2	2	4.04
Re	<10	<2.8	3	<4.08

*See *Table 4-1* for calculation of e/a ratio in terms of ΔGN.

recent studies of martensitic transformation in representative members of the Ti-V, Ti-Nb, and Ti-Mo systems by FLOWER et al. [FLO82] are in good agreement with this, α'' having been detected in Ti-10.6V(10.0 at.%),* Ti-14Nb(8 at.%), Ti-20Nb(11 at.%), and Ti-4Mo through Ti-8Mo (2 through 4 at.%). Not all authors agree that the higher concentration Ti-V martensites are orthorhombic. WILLIAMS had not succeeded in detecting the α'' variant in Ti-11.6V (11.0 at.%) and attributed its presence, apparent or otherwise, in earlier experiments to difficulties encountered in the handling and study of thin foils in general [WIL73]. In Ti-V and the like, such difficulties are of course further exacerbated by the instabilities of the lattices under study.

Since stress-induced martensite can occur only in alloys not already transformed by quenching, it is obviously confined to the higher concentration ranges; in the light of the foregoing observations on the structures of quenched martensites it is, therefore, not at all surprising to find that deformation martensite is invariably of the orthorhombic variety [WIL73].

4.2.4 Crystallographic, Thermodynamic, and Acoustic Aspects of the Martensitic Transformation

(a) Crystallography. Details of the crystallographic $\beta \rightarrow \alpha^m$ transformation in Ti alloys have been reviewed and discussed by several authorities, notably OTTE [OTT70], HAMMOND and KELLY [HAM70], and WILLIAMS [WIL73]. The crystallographic approach focuses attention on the influence of various possible shear systems on the habit plane of the

---

*The spinodal decomposition of aged, previously martensitic, Ti-10.6V is good indirect evidence that its structure was originally α'' and not α' [FLO82] — see Sect. 4.2.5(b).

transformed plates with the hope of determining which of them are consistent with the results of habit-plane orientation measurements and, consequently, which dislocations dominate the transformation. By a process of elimination, the justification for which forms the basis for discussion and is the essence of the exercise, a multitude of possible dislocation paths are reduced to just a few. According to OTTE [OTT70] the entire $\beta \to \alpha'$ transformation process was reducible to an activation of the shear systems:

$$[111]_\beta (11\bar{2})_\beta \equiv [2\bar{1}\bar{1}3]_{\alpha'}(\bar{2}112)_{\alpha'}$$

and

$$\left.\begin{array}{c} \\ [111]_\beta (\bar{1}01)_\beta \equiv [2\bar{1}\bar{1}3]_{\alpha'}(\bar{1}011)_{\alpha'} \end{array}\right\} \quad (4\text{-}1)$$

It has been well established that the transformation from β to α' is characterized by a habit plane near $\{334\}_\beta$ [BLA70][WIL73][SHI77] and that the Burgers orientation relation is closely obeyed [WIL73][SHI77][DAV79] [FLO82]; stated in crystallographic terms, this implies: $(0001)_{\alpha'}\|(011)_\beta$; $\langle 11\bar{2}0\rangle_{\alpha'} \sim \|\langle 11\bar{1}\rangle_\beta$. The results of the study by DAVIS *et al.* [DAV79] of Ti and the alloys Ti-Mo(2, 4, 6, and 8 wt.%) were entirely consistent with that general statement. Furthermore they were able to show, for Ti and all four of the alloys, that the transformation of β to α' using the Burgers relationship can be achieved via:

a 10% contraction along $[100]_\beta$ which corresponds to $[2\bar{1}\bar{1}0]_{\alpha'}$,

a 10% expansion along $[01\bar{1}]_\beta$ which corresponds to $[01\bar{1}0]_{\alpha'}$,

a 1% expansion along $[011]_\beta$ which corresponds to $[0001]_{\alpha'}$.

The crystallography of the orthorhombic α'', as formed in quenched Ti-Nb(20 at.%), has been discussed by HATT and RIVLIN [HAT68], who concluded that the following orientation relationships existed (accuracy $\pm 0.5°$):

$[100]_{\alpha''}$ 2° from $\langle 001\rangle_\beta$,

$[010]_{\alpha''}$ 2° from $\langle 110\rangle_\beta$,

$[001]_{\alpha''}\|\langle 1\bar{1}0\rangle_\beta$.

(b) Thermodynamics and Acoustics. Underlying the phenomenological crystallographic theory of the martensitic transformation are several levels of semimechanical and, finally, mechanistic interpretations. Earlier in this section it was shown that similar e/a ratios characterized the limit of bcc stability with respect to either the ω-phase or martensitic transformation. The underlying common ingredient is a localized soft-phonon mode whose relationship to those two types of transformation can be traced through a series of papers by DE FONTAINE et al. [DEF70, DEF71], COOK [COO73, COO74, COO75ª], CLAPP [CLA73], and SUZUKI and WITTIG [SUZ75]. In the last two papers a formalism was developed for comparing martensitic transformation to spinodal decomposition, which in turn has been coupled by others to ω-phase precipitation. A formal identification is achieved by comparing, in macroscopic terms, composition fluctuations to displacement fluctuations (i.e., deformation). Thus, just as spinodal decomposition, with which the names of HILLERT [HIL61] and CAHN [CAH61] are associated, can be treated as a free energy, g, instability with respect to a fluctuation in composition, c, the threshold of which is defined by $(\partial^2 g/\partial c^2) = 0$, the martensitic transformation, according to CLAPP [CLA73] and SUZUKI et al. [SUZ75], can be triggered by strain, ϵ, fluctuations in the vicinity of a strain spinodal, $(\partial^2 g/\partial \epsilon^2) < 0$. The approach taken by CLAPP to describe the initial nucleation of the martensitic transformation commenced with an expression for the free-energy density as a function of strain to third order in the elastic constants. For a cubic lattice, the vanishing of one or more elements of a 6×6 matrix representing the second derivatives of g with respect to strain, viz:

$$\partial^2 g/\partial \epsilon_I \partial \epsilon_J \equiv g_{IJ}(\epsilon) \quad (I, J = 1, \ldots, 6) \; , \qquad (4\text{-}2)$$

defines a strain spinodal for the corresponding directions in strain space. In lattices which are prone to transform, g_{IJ}, already small, may be forced to zero by the application of a few percent strain (the deformation-martensite effect). The condition for static instability may also be approached via the introduction of lattice imperfections such as point and line defects, and even the surface itself, which may place the crystal close to one or other of its strain spinodals; an appropriately applied dynamic strain, in the form of one half cycle of a lattice vibration, will then be able to nucleate a burst of transformation.

The discussion of the previous section led to the conclusion that the ω-phase and martensitic transformations are closely related. A connection between them in terms of soft-mode lattice dynamics is now apparent. As we shall see, the DE FONTAINE theory [DEF70, DEF71] of the ω phase required the bcc lattice to be unstable with respect to longitudinal phonons centered about $\frac{2}{3}\langle 111 \rangle$.

4.2.5 Transformations from and to the β Phase: Case Studies of Some Representative Ti-Base Alloy Systems

In α-stabilized alloys the M_s temperature increases with solute content; in β-phase alloys it decreases. Thus, if the transformation has a diffusional component (temperature dependent), it will respond differently to increases in solute concentration in these two classes of systems. The martensitic transformation is characterized phenomenologically by the assignment of several temperatures: M_s, the start of the martensitic transformation during quenching, and M_f, its finish; β_s, the start of the $\alpha^m \to \beta$ reversion on up-quenching; T_0, the temperature at which α^m and β are in thermodynamic equilibrium (i.e., have the same free energies).

(a) The Titanium-Aluminum System. The martensitic transformation in Ti-Al and the effect of cooling rate on it have been studied by JEPSON et al. [JEP70]. The various transformation temperatures encountered are listed in *Table 4-5*. For this system it was found impossible to suppress the $\beta \to \alpha'$ transformation at the cooling rates available, viz, $\gtrsim 2 \times 10^5 \,°\text{C s}^{-1}$. If the martensitic transformations were truly athermal, there would be no change in M_s with changing cooling rate. The fact that M_s did begin to drop at cooling rates in the vicinity of $5 \times 10^4 \,°\text{C s}^{-1}$ indicated that the transformation was thermally activated. The closeness of M_s to β_s (*Table 4-5*) indicated that the driving force for the transformation was small.

(b) The Titanium-Molybdenum System. As indicated earlier within the context of structural alloys, Ti-Mo is an important prototype system, establishing as it does the basis of several commercially interesting multicomponent formulations [OHT73], particularly β-III. Ti-Mo alloys have been the subjects of several recent investigations. DAVIS, FLOWER, and WEST studied the martensitic transformation itself [DAV79] as well as the decomposition of the Ti-Mo martensites during aging [DAV79ᵃ][FLO82],

Table 4-5. Phase-Transformation and Equilibrium-Phase-Boundary Temperatures for Ti-Al Alloys [JEP70]

Al content, at.%	Transformation temperatures,* °C			Phase-boundary temperatures, °C	
	T_0	M_s	β_s	$\alpha/(\alpha+\beta)$	$(\alpha+\beta)/\beta$
5	943	918	940	925	960
10	1008	960	1010	975	1040
15	1066	1015	1080	1035	1100
20	1113	1060	1110	1080	1100
					(1158)

*T_0 = Temperature corresponding to zero free-energy difference between the α' and β phases (calculated). M_s = Martensite start ($\beta \to \alpha'$) temperature. β_s = Martensite reversion ($\alpha' \to \beta$) temperature.

while crystallographic and microstructural studies of the stress-induced transformations have been conducted by OKA and TANIGUCHI [OKA80] and HIDA *et al.* [HID80], respectively.

The Mo concentrations studied by FLOWER *et al.* [FLO82] were 2, 4, 6, and 8 wt.% Mo (1, 2, and 4 at.%) and thereby spanned the composition, ~4 wt.% Mo (~2 at.% Mo), at which the quenched structure changed from α' to α'' [BAG59]. With increasing Mo content there is a transition in the martensite morphology from massive to acicular (Sect. 4.2.2). Ti-2Mo was already acicular with only a small proportion of the massive; the higher concentration alloys were entirely acicular. In agreement with BAGARIATSKII *et al.* [BAG59], DAVIS *et al.* [DAV79] showed that, whereas the martensitic form of Ti-4Mo was hexagonal α', that of Ti-6Mo was orthorhombic α''. They also mentioned, with reference to earlier work, that whereas α'' was twinned, the α' was generally dislocated; but they modified the strength of this comparison by pointing out that twinning was not completely confined to the α'' variant [FLO82].

Layers of retained β were detected in Ti-2Mo and Ti-4Mo, but not in Ti-6Mo and Ti-8Mo. This observation supports the general conclusion that in dilute β-stabilized Ti-TM alloys, with their relatively high M_s temperatures, water quenching is insufficiently rapid to completely inhibit diffusional reactions. Some segregation of the Mo takes place, leading to the production of some untransformable β phase. It was suggested by DAVIS *et al.* [DAV79] that the existence of the diffusional component may have been a factor in the formation of massive martensite. As the solute concentration increases, M_s decreases, the diffusional contribution becomes suppressed, and a full transformation to α^m is able to take place. This continues until M_f drops below the temperature of the quench bath, again enabling the retention of some β phase.

The results of the aging studies of FLOWER *et al.* [FLO82] were particularly interesting. It is not appropriate to describe them in detail here; however, it is useful simply to mention that, whereas α' reverts to β via a nucleation-and-growth process, the initial stage of α'' decomposition may be spinodal, to $\alpha''_{lean} + \alpha''_{rich}$, thereby yielding a characteristically modulated microstructure. Thus through observing the microstructure of the aged product, the nature of the original as-quenched structure can in some cases be deduced.

As indicated in Sect. 4.2.2, the application of stress to a metastable β-Ti-TM alloy will result in either martensitic transformation or twinning. The composition ranges of these two transformation products have been investigated by OKA and TANIGUCHI [OKA80] using a series of Ti-Mo alloys of compositions: 9.3, 10.5, 11, 13, 14, and 15.5 wt.% Mo. Using x-ray diffraction and TEM they were able to demonstrate that the deformation of Ti-Mo(9~11 wt.%) led to stress-induced α'', while that of

Ti-Mo(11~15.5 wt.%) resulted in {332} twinning in agreement with numerous earlier reported results.

(c) The Titanium-Niobium System. (*i*) *Ranges of Occurrence of the Martensitic α' and α'' Phases.* For Ti-Nb, an upper-concentration limit for α' martensite of 6 at.% (10.5 wt.%) has been reported by BAGARIATSKII *et al.* [BAG59], *Table 4-4*. The limit of 3 at.% Nb referred to by HATT and RIVLIN [HAT68] and perpetuated by some subsequent authors may have originated in what seems to be a plotting error in Fig. 1(d) of [BAG59], although it had subsequently been corrected, by implication, in that same paper. At concentrations higher than ~11 at.% Nb, the quenched-martensitic structure is orthorhombic α'', which seems to persist, more or less, out to concentrations as high as 25 at.% Nb depending on the quenching conditions [HIC69a]. Indeed, the M_s transus of *Fig. 4-9* intersects the 300-K axis at 26.5 at.% Nb. However, HICKMAN's rapidly quenched Ti-Nb(27 at.%) was single-phase β [HIC69a]. In other studies related to the high-concentration limit of α'', BALCERZAK and SASS [BAL72] showed that the structure of Ti-Nb(18.4 at.%), oil-quenched from 900°C, was α'', as was that of Ti-Nb(20 at.%), water-quenched from 900°C, according to BAKER and SUTTON [BAK69]. With Ti-Nb(20.7 at.%), HATT

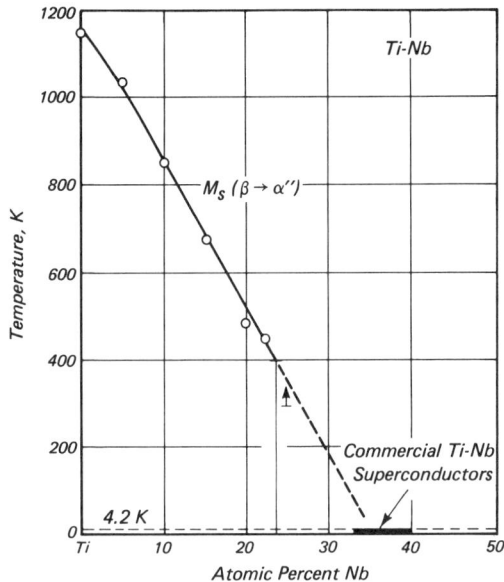

Fig. 4-9. The martensitic $\beta \to \alpha''$ transformation curve for Ti-Nb based on the data of DUWEZ [DUW53], BROWN *et al.* [BRO64], and BAKER [BAK71]. The extrapolation to liquid-He temperatures was suggested and discussed by KOCH and EASTON [KOC77].

and RIVLIN [HAT68] showed that water quenching from 900°C yielded α'' plus a trace of β. Ti-Nb(22 at.%), gas-quenched at about 10^{3}°C s^{-1} from 1200~1300°C, yielded $\alpha'' + \beta$ [HIC69^a], as did Ti-Nb(22.6 at.%) oil-quenched from 800°C [BAL72] and gas-quenched Ti-Nb(25 at.%) [HIC69^a]. Ti-Nb(25.6 at.%) oil-quenched from 900°C was essentially $\omega + \beta$ [BAL72].

(*ii*) *Morphology of the Ti-Nb Martensite.* In water-quenched Ti-0.5Nb $(0.2_6$ at.%), the α' martensite was predominately massive, while that in Ti-1.0Nb $(0.5_2$ at.%) was predominately acicular [FLO82]. In both cases layers of retained β were present, an expected consequence of water-quenching through the high-temperature segment of M_s. Both the occurrence and the microstructure of the quenched martensite are sensitive to the rate of cooling. The most detailed study of the effect of cooling rate was undertaken by JEPSON, BROWN, and GRAY [JEP70]. In experiments in which the cooling/quench rate was varied from 0.2°C s^{-1} to 2×10^5°C s^{-1}, considerable control over the microstructure of the product was able to be exerted. For example, the structure of Ti-Nb(17.5 at.%) quenched at 1.0×10^5°C s^{-1} was equiaxed single-phase β. Naturally in less concentrated alloys, martensitic transformation was able to take place at quench rates fast enough to retain the β phase in the more concentrated ones. JEPSON *et al.* made the interesting discovery that the effect of increasing the Nb content was not only to decrease M_s but also to make the $\beta \rightarrow \alpha^m$ transformation more sluggish. As with the Ti-Al alloys referred to above, the transformation temperatures in Ti-Nb(0 through 17.5 at.%) all decreased with increasing quench rates, especially at high rates, indicating that the transformations were actually thermally activated [JEP70].

(*iii*) *Reversion and Aging of the Ti-Nb Martensites.* Numerous authors have investigated the thermal stabilities and modes of decomposition of tempered (aged) martensites. Using a thermal-arrest technique, JEPSON *et al.* [JEP70] have determined both the M_s temperature and the temperature, β_s, at which the *reverse* transformation commences. The results are given in *Table 4-6.* If M_s and β_s are fairly close, the temperature T_0, corresponding to equality of the β and α^m free energies, can be taken as their mean. Other workers, notably HATT and RIVLIN [HAT68], BAKER and SUTTON [BAK69], HICKMAN [HIC69^a], and FLOWER *et al.* [DAV79^a][FLO82], have considered the manner in which the α^m decomposes and the natures of the products, which may be β, $\omega + \beta$, or $\alpha + \beta$, depending on the conditions. The situation can be loosely, but instructively, summarized with the aid of *Fig. 4-10*, a heuristic nonequilibrium/equilibrium phase diagram illustrating the relative positions of the $M_s^{\alpha'}$ and $M_s^{\alpha''}$ transi (after [FLO82]) and the isothermal $\omega + \beta$ phase region. Several classes of decomposition take place, depending on the composition of the Ti-Nb martensite and the temperature. Upon aging the α' martensites ($\gtrsim$11 wt.%, 6 at.% Nb), the

**Table 4-6. Phase-Transformation and Equilibrium-Phase-Boundary
Temperatures for Ti-Nb Alloys**

Nb content, at.%	Transformation temperatures,* °C			$\beta/(\alpha+\beta)$ Boundary, °C	
	M_s				
	[Jep70]	[Duw53]	β_s [Jep70]	[Jep70]	[Han51]
0	855	855	—	885	885
$2\frac{1}{2}$	753	—	—	—	—
5	720	760	—	760	810
$7\frac{1}{2}$	619	—	646	—	—
9	567	—	592	—	—
10	560	600	540	650	765
11	517	—	530	—	—
$12\frac{1}{2}$	455	500	455	620	740
15	385	400	387	585	725
$17\frac{1}{2}$	300	—	317	545	705

*See *Table 4-5* for definitions.

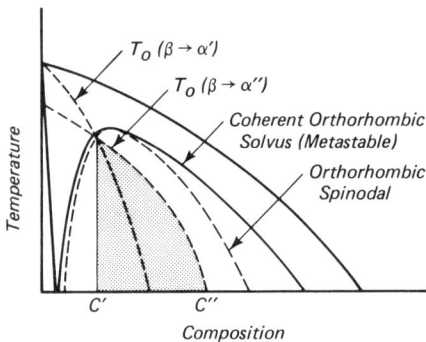

Fig. 4-10. Schematic representation of phase relationships in β-isomorphous Ti-TM alloys (i.e., Ti-V, Ti-Nb, and Ti-Mo) for the purpose of illustrating the modes of martensitic transformation and the decomposition of martensites by spinodal decomposition [Flo82], see also [Dav79a].

equilibrium β phase nucleates heterogeneously. As indicated in *Fig. 4-10*, the α'' martensites within a certain composition range are able to spinodally decompose into $\alpha''_{lean} + \alpha''_{rich}$ [Flo82][Dav79a]. This process, which has been observed to take place in Ti-Nb(14-20 wt.%)(8-11 at.% Nb) [Flo82] begins during the quench, continues on further aging, and eventually proceeds to $\alpha+\beta$. Alloys sufficiently high in Nb content and aged at moderate temperatures revert to metastable β, which then decomposes into $\omega+\beta$. This process has been noted in water-quenched α''-Ti-Nb(20 at.%) aged at 330°C [Bak69], water-quenched α''-Ti-Nb(20.7 at.%) aged at 335°C [Hat68], Ti-Nb(22 at.%) aged above 200°C [Hic69a], and Ti-Nb(25 at.%) aged above 150°C [Hic69a], both of the latter alloys having been previously gas-quenched.

(d) Quenched Multicomponent $\alpha + \beta$ Alloys. (i) *Ti-6Al-4V*. The various types of structures encountered in Ti-6Al-4V can be readily appreciated with the aid of *Fig. 3-14*. If the alloy is cooled rapidly from the bcc field, it will undergo a transformation to acicular α (referred to then as transformed β, β_{tr}) en route. The prior-β grain boundaries, decorated by α phase, will still be visible in the β_{tr} structure, where they demarcate the junctions between the rafts of α-phase needles.* If the alloy is equilibrated high in the $\alpha + \beta$ region, it will contain a large volume-fraction of β phase of a concentration sufficiently low that it will again transform on rapid cooling. The lower the temperature of $\alpha + \beta$ equilibration, the smaller the volume-fraction of β phase but the higher its V concentration and, consequently, its β stability. Following such a heat treatment it is possible to obtain structures consisting of large equiaxed α grains with small amounts of retained β preserved intergranularly. All these features were depicted in *Fig. 3-17*.

(ii) *Ti-6Al-2Sn-4Zr-2Mo*. The transformation kinetics of Ti-6242 have been studied by conventional quench techniques to produce T-T-T diagrams such as that depicted in *Fig. 4-11*. Although full martensitic transformation is, in principle, possible during quenching from above the β transus (990°C), transformation of the β phase to Widmanstätten $\alpha + \beta$ (occasionally designated $(\alpha + \beta)_W$) by nucleation and growth of α platelets takes place very rapidly. For example, α-phase precipitation will commence after the alloy has been held for a little more than 1 sec at 800°C. The properties of $(\alpha + \beta)_W$, formed in a test sample of Ti-6242-0.1Si (condition: 2h/1024°C/air cool) in response to deformation and thermal cycling, have been examined in detail by GEGEL and colleagues [GEG80^a]. They were able to show that the transformed-β structure (designated β_{tr} or

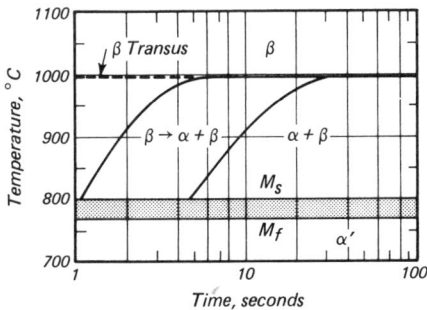

Fig. 4-11. Isothermal T-T-T diagram for Ti-6Al-2Sn-4Zr-2Mo [Woo72; p. 1-6:72-3].

---

*In alloys which contain a substantial concentration of Al, such as Ti-6Al-4V, α phase may nucleate directly from the metastable β in the form of plates or needles (acicular particles), aligned parallel to $\{110\}_\beta$, yielding a "basket-weave"-type of structure. The Widmanstätten-α plates, as they are called, are separated by a β-phase matrix [WIL73].

$(\alpha+\beta)_W)$ possessed several interesting and important features: (i) a well-defined "interface phase" separating the α and β components of the Widmanstätten structure; (ii) a characteristic distribution of solute species across α/interface/β, see *Fig. 3-21*; and (iii) a variation of the composition-dependent e/a ratio across these three phases and the possible occurrence within one of them of a vanishing e/a-controlled elastic shear modulus C' (see Sects. 3.1.4, 4.1.3, and 5.7).

The STEM-analyzed compositions of the α, β, and interface phases, as measured by GEGEL and colleagues, are given in *Fig. 3-21* and *Table 3-2*. In order to translate composition into electron/atom ratio, use is made of the general formula: $(e/a) = 4.00 + \sum\limits_{i}^{n} f_i \Delta(e/a)_i$, where f_i is the mole (atomic) fraction of the ith solute and $\Delta(e/a)_i$ is the difference between its e/a ratio and that of Ti ($= 4.00$). For the phases and components listed in *Table 3-2*, then:

$$(e/a)_\alpha = 4.00 - (0.08 \times 1) + (0.005 \times 2) = 3.93 \ ,$$

and

$$(e/a)_\beta = 4.00 - (0.01 \times 1) + (0.08 \times 2) = 4.15 \ .$$

Viewing this value for $(e/a)_\beta$ in the light of *Table 4-1*, and the discussions of Sects. 3.1.4 and 4.1.3, it is clear that the β phase separating the Widmanstätten platelets in the structure of quenched Ti-6242 (or Ti-6242-0.1Si) is just on the threshold of β stability. Furthermore, since the Mo concentration decreases towards the edges of the β layer, its shear modulus, C' (see *Fig. 3-3*), is likely to vanish close to the interface, thus creating a condition for mechanical instability there. The stability of the interface layer itself is not at issue since its structure, if analogous to that formed between the α and β phases in transformed Ti-6Al-4V [WIL82[b]], is probably α rather than β.

4.3 Occurrence and Structure of Quenched (Athermal) ω Phase

4.3.1 Occurrence and Morphology

The development of our current understanding of ω-phase precipitation in Ti-base and Zr-base alloys can be traced with the aid of *Fig. 4-12*, a literature survey in diagrammatic form. During the 1950s, ω-phase precipitation was shown to occur athermally over a very narrow composition range during the rapid quenching from the elevated-temperature β field of numerous Ti-TM alloys. Athermal ω phase is a diffusionless transformation product and as such cannot be suppressed no matter how rapid the

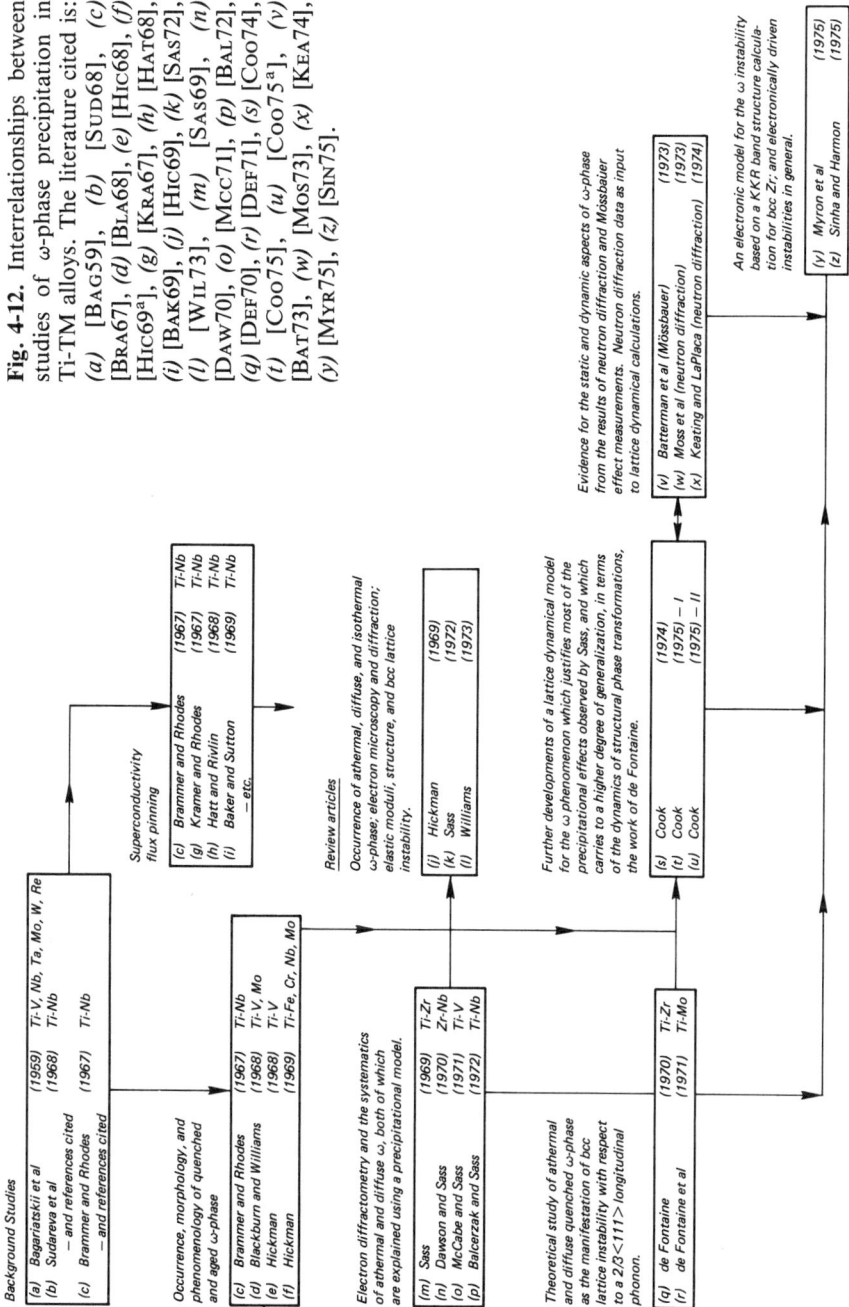

Fig. 4-12. Interrelationships between studies of ω-phase precipitation in Ti-TM alloys. The literature cited is: (a) [BAG59], (b) [SUD68], (c) [BRA67], (d) [BLA68], (e) [HIC68], (f) [HIC69a], (g) [KRA67], (h) [HAT68], (i) [BAK69], (j) [HIC69], (k) [SAS72], (l) [WIL73], (m) [SAS69], (n) [DAW70], (o) [MCC71], (p) [BAL72], (q) [DEF70], (r) [DEF71], (s) [COO74], (t) [COO75], (u) [COO75a], (v) [BAT73], (w) [MOS73], (x) [SIN75], (y) [MYR75], (z) [KEA74].

quench rate (at least up to $1.1 \times 10^4 \,°C \, s^{-1}$) [DUE80]. Early x-ray studies showed its structure to be either hexagonal (SILCOCK [SIL58]) or trigonal (BAGARIATSKII et al. [BAG59]), depending (as a result of more recent work) on solute concentration. The compositions of athermal ω phase in Ti-V, -Nb, -Cr, -Mo, -W, -Mn, and -Re, as determined by BAGARIATSKII et al. [BAG59] on quenched alloys using x-ray techniques, are presented in *Table 4-2(a)*. They can be seen to be in excellent agreement with results obtained a decade later by HICKMAN [HIC69[a]] in his studies of the constitutions of aged alloys. GUSEVA and DOLINSKAYA [GUS82], applying x-ray techniques to quenched alloys, extended the range of materials first investigated by BAGARIATSKII et al. [BAG59]. The results of this series of measurements, which included an investigation of some ternary alloy systems, are presented in *Table 4-2(b)*.

In a series of investigations using single-crystal x-ray techniques, HICKMAN [HIC68, HIC69, HIC69[a]] had studied the occurrence of ω phase in the systems Ti-V, -Cr, -Fe, -Co, and -Ni, paying particular attention to the composition range over which it appeared during quenching, and to the effects of isothermal aging which led eventually to the establishment of an "equilibrium" $\omega + \beta$ state. During this period of renewed interest in ω phase between about 1968 and 1972, the use of transmission electron microscopy (TEM) and electron diffraction led rapidly through a period of phenomenological exploration, with which the names of BRAMMER and RHODES (Ti-Nb, [BRA67]), BLACKBURN and WILLIAMS (Ti-V, Ti-Mo [BLA68]), and KRAMER and RHODES (Ti-Nb, [KRA67]) are associated, to an elegant series of quantitative investigations by SASS and his students of the systems Ti-Zr [SAS69], Zr-Nb [DAW70], Ti-V [MCC71], and Ti-Nb [BAL72], which together with an important investigation of Ti-Mo by DE FONTAINE, PATON, and WILLIAMS [DEF71] laid the groundwork for our current understanding of athermal and "diffuse" quenched ω phase, and indeed the entire ω-phase precipitation phenomenon.

The morphology of the ω phase has been investigated by BLACKBURN and WILLIAMS [BLA68, BLA70], who have shown that it forms as cubes when in the Ti-(3d)TM alloys Ti-V, Ti-Cr, Ti-Mn, and Ti-Fe and as ellipsoids in the Ti-(4d)TM alloys Ti-Nb and Ti-Mo. The precipitate morphology is related to the Ti/TM atomic-volume ratios, hence to lattice misfit. Thus, it turns out that when the misfit is low ($\approx 0.5\%$), the precipitate is ellipsoidal [BLA68][HIC69[a]]; otherwise, it is cubic.

4.3.2 Crystallography and Lattice Dynamics

It has been well established that the ω phase is related to the bcc parent crystal according to $(0001)_\omega \| (111)_\beta$; $[2\bar{1}\bar{1}0]_\omega \| [1\bar{1}0]_\beta$ implying the existence of four variants of it with respect to the bcc lattice [SIL58][BAG58]

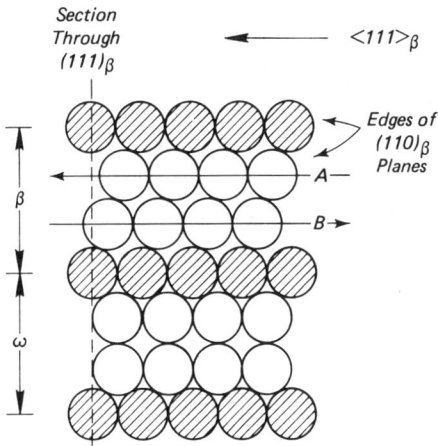

Fig. 4-13. "Linear-fault" model for ω-phase transformation: $(101)_\beta$ section (in plane of page) through bcc crystal depicting the transformation to ω phase as a displacement of adjacent $(110)_\beta$ planes (say, planes A and B) [BAG59].

[BAL72]. ω phase may be regarded as being developed within that lattice by applying to pairs of adjacent $(110)_\beta$ planes equal and opposite shears, in the $\langle 111 \rangle_\beta$ direction, through distances about equal to $\frac{1}{6}$ of the separation of the $(111)_\beta$ planes. The arrows in *Fig. 4-13* indicate the planes, or rows, of atoms involved and the directions of the shears required. The coherent β/ω interface at the boundary $(110)_\beta$ plane is an important feature. If z is the separation of the $(111)_\beta$ planes $(= a\sqrt{3}/2$, where a is the bcc lattice parameter), then: (i) displacements of the A and B atoms by the amounts $\pm(\frac{1}{6})z$ lead to the hexagonal structure of *Fig. 4-13* proposed originally by SILCOCK [SIL58] as a result of measurements on a low solute-content ω phase; and (ii) displacements of $\pm 0.15z$ yield the trigonal structure originally proposed by BAGARIATSKII (see [BAG59]) and now understood to be characteristic of higher solute-content ω phases.

Assuming touching hard spheres of constant diameter (ϕ) in the simple geometrical model of *Fig. 4-13*, the separation of the $(110)_\beta$ planes, originally $2\sqrt{2}\,\phi$, shrinks to $(1 + \sqrt{3})\phi$ upon transformation to ω phase, a 3% contraction which justifies the tendency for ω to form in Ti and Zr under pressure [JAY63]. Adoption of this conventional crystallographic approach enabled most of the gross features of quenched ω phase, as determined by TEM and electron diffraction [SAS69][DAW70][MCC71][BAL72], to be interpreted. *Fig. 4-13* suggests a $\langle 111 \rangle$ texture to the ω precipitation. Indeed, the model proposed by SASS *et al.* [DAW70][MCC71][BAL72] was based upon $\langle 111 \rangle$ rows of particles, 10 to 15 Å in diameter and 15 to 25 Å apart. According to that static-particle model, athermal ω phase, whose electron diffractograms, *Fig. 4-14*, are characterized by sharp spots and straight "lines of intensity," was made up of clusters of such rows, while the broad reflections and the either straight or curved lines of intensity

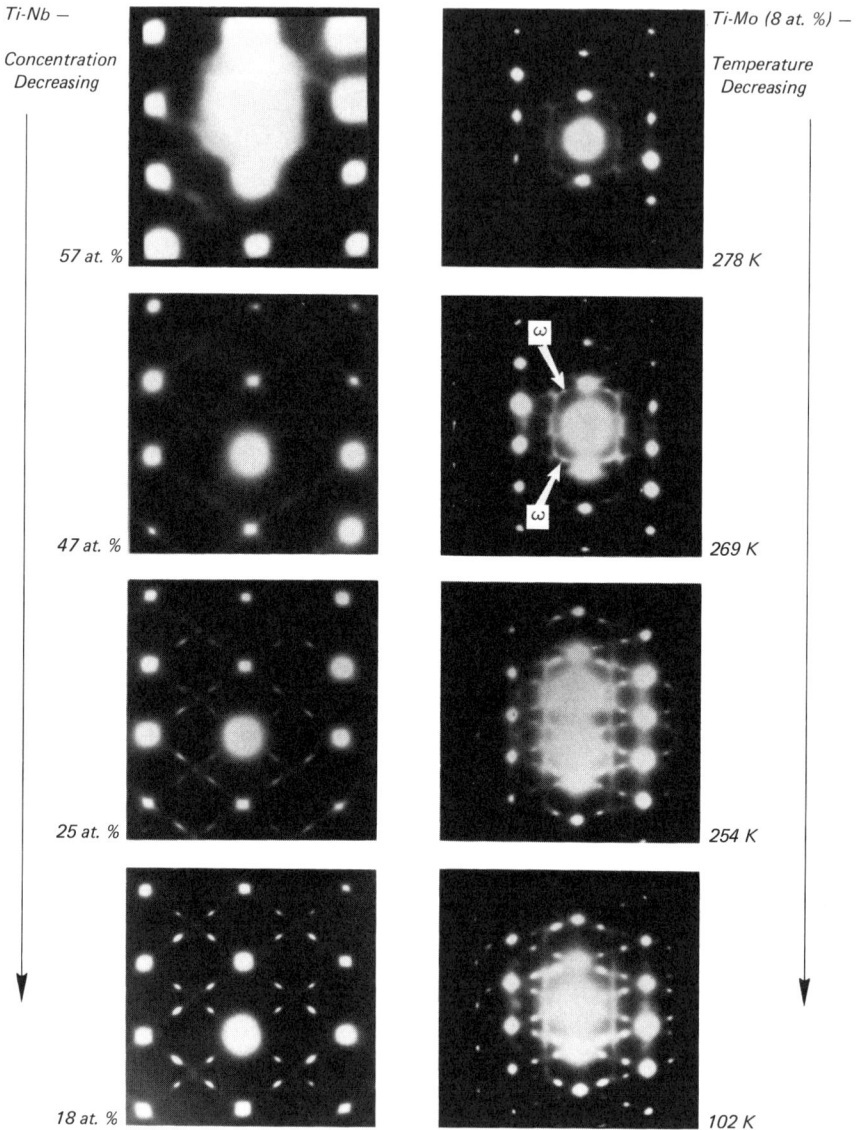

Fig. 4-14. Changes from diffuse to sharp ω reflections from quenched Ti-TM alloys in response to either decrease in the solute content or decrease in the temperature. *Left side*: As-quenched Ti-Nb alloys in the (110) reciprocal-lattice section [BAL71, BAL72][SAS72]; photographs courtesy of S.L. Sass (Cornell University). *Right side*: As-quenched Ti-Mo(8 at.%) in the (131) reciprocal-lattice section [DEF71]; photographs courtesy of J.C. Williams (Carnegie-Mellon University).

("diffuse streaking") of "diffuse ω" were supposed to originate from either individual rows of particles, or isolated particles, respectively.

But the reversible nature of ω precipitation under temperature cycling between room temperature and 100 K calls for more than a static crystallographic interpretation. *Fig. 4-14* compares the composition dependence of the electron diffractograms of Ti-Nb with the temperature dependence of those of Ti-Mo. It seems that the effect of lowering temperature is similar to that of lowering solute concentration, in that in both cases curvilinear lines of diffuse intensity become straight and well defined. The figure serves to demonstrate, moreover, the reason for the uncertainties and arguments which have been associated with the assignment of compositional limits for athermal-ω-phase formation. As pointed out by WILLIAMS [WIL73]: (i) since the diffuse streaking tends to coincide with the positions of the ω-particle reflections when they are present, there is no sharp line of demarcation separating the regions of athermal and diffuse ω; and (ii) the reversibility of the ω makes the specification of temperature particularly important, especially when relating structure to low-temperature physical properties such as the superconducting transition temperature. The soft-phonon mechanistic model of the ω-phase effect, originating with the work of DE FONTAINE [DEF70] and developed more fully in the paper by DE FONTAINE, PATON, and WILLIAMS [DEF71], provided a satisfactory rationalization, in lattice-dynamical terms, for both the temperature- and composition-dependences of the athermal and diffuse ω phases.

The dynamical equivalent of the linear-fault crystallographic model centers about the proposed existence of a longitudinal phonon propagating in the $\langle 111 \rangle$ direction. The wavelength necessary to achieve (with the aid of anharmonicity) the necessary displacements of the A and B atoms (*Fig. 4-13*) is illustrated in *Fig. 4-15*, which represents a unit cell of the earlier figure. Clearly, if the shaded atoms are to remain unmoved, while A and B are to be shifted in opposite directions as shown, a longitudinal wave of wavelength equal to the separation of the $(111)_\beta$ planes is needed, viz, a longitudinal phonon with wave-vector $\frac{2}{3}\langle 111 \rangle$. It is the instability of the bcc lattice to this disturbance that is responsible for the athermal transition. If the crystal is removed from a region of instability either by variation of its composition or its temperature, the $\frac{2}{3}\langle 111 \rangle$ phonon is responsible for the so-called reciprocal-lattice "streaking" effect also known as "diffuse" ω, *Fig. 4-14*. With particular reference to the Ti-V system, *Table 4-7* summarizes and compares the lattice-dynamical and linear-fault approaches as they would apply to quenched ω phases in Ti-TM alloys in general. Both philosophies are in agreement in the athermal-precipitate regime in that they represent, respectively, "cause" and "effect". They do not, however, agree in the diffuse regime: since diffuse-ω domains cannot

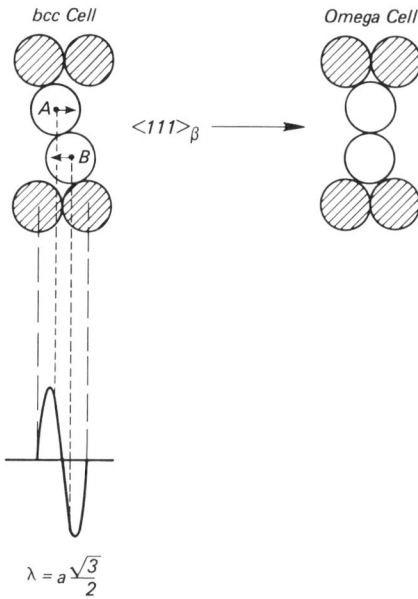

Fig. 4-15. Schematic representation of the bcc → ω transformation induced by the application of a $\frac{2}{3} \langle 111 \rangle$ longitudinal displacement wave to the bcc lattice [Sas72].

be imaged in the electron microscope, the linear-fault model is incapable of explaining the curvilinear diffraction lines of intensity. The phonon model does so by invoking the coupling of displacement modes [Def71] — the combined process being represented by a hyperbola asymptotic to the straight lines of the independent processes.

Further developments and generalizations of the lattice-dynamical model for phase stability, with particular reference to the quenched ω-phase phenomenon, have been made by Cook [Coo75, Coo75ᵃ] with help from the observations of Sass and de Fontaine, and the results of the neutron diffraction studies of Zr-Nb alloys by Moss, Keating, and Axe [Mos73][Kea74]. Lattice fluctuation effects in the region of diffuse ω phase have been probed using both neutron diffraction [Mos73][Kea74] and Mössbauer-effect measurements [Bat73]. There seems little doubt that the $\frac{2}{3}\langle 111 \rangle$ soft mode, already present in the pure "solute" elements, interacting with a lattice of temperature- and composition-dependent relative stability, is responsible not only for the athermal precipitate but also for the diffuse ω, which represents to varying degrees dynamical fluctuations between the crystalline bcc and ω phases.

It is important to remember that even the lattice-dynamical model is phenomenological, in the sense that a virtual-crystal approximation (identical atoms) is assumed and the electronic properties are disguised as force constants. A more fundamental understanding of the transformation requires an examination of what is generally referred to as the "electronic

Table 4-7. Electron Diffraction and Lattice-Dynamical Approaches to ω-Phase Formation in Group-IV-Base Binary Transition-Metal Alloys

The Ti-V system has been selected as an example. The headings of the last two columns of the table indicate consecutively: technique, author, class of alloy, approach.

Typical solute concentration ranges in Ti-V	Diffraction effects as seen in electron microscopic studies of quenched alloys [SAS72]	• Electron microscopy • S.L. SASS [SAS72] • Real alloy* • Static*	• Lattice-dynamical calculations • D. DE FONTAINE [DEF70, DEF71] • Virtual crystal • Dynamical*
13 at.% V	Sharp ω reflections and straight lines of intensity	Clusters of rows of particles plus individual rows	Propagating (long-range) $\frac{2}{3}\langle 111 \rangle$ longitudinal mode of wavelength $(\sqrt{3}/2)$ times (bcc lattice parameter)
13,15, and 19 at.% V	Broad ω reflections and straight lines of intensity	Rows of 20-Å particles; spaced 22 Å, typically 5 to a row	$\frac{2}{3}\langle 111 \rangle$ vibrational modes tending to become more localized and less coherent
25 at.% V	Broad ω reflections, lines of intensity curved	Isolated 15-Å particles	
25-55 at.% V	Weak, broad lines of intensity, becoming diffuse and circular as solute concentration increases	Not interpretable in terms of precipitation	Curvilinear diffuse reflections due to pronounced coupling between local vibrational modes

*These two columns are coupled by the following: (i) the dynamical effect yields a static (observable) atomic displacement through anharmonicity; (ii) the ω phase becomes "frozen in" as a result of solute-solvent interdiffusion in a real alloy.

structure" of the alloy. It is neither possible nor appropriate to deal with the problem in this space. Suffice it to mention that SINHA and HARMON [SIN75] have constructed a dielectric screening model for the treatment of general so-called "electronically driven" instabilities, while MYRON, FREEMAN, and MOSS [MYR75] have, with the aid of KKR calculations of the band structure and Fermi surface of bcc Zr, and some assumptions regarding the perturbation of this structure during the addition of Nb, directly addressed the subject of the ω instability.

4.3.3 ω-Phase Precipitation in Binary and Multicomponent Alloys: Case Studies of Some Representative Ti-Base Alloy Systems

(a) The Ti-Nb System. The occurrence and structure of the athermal ω phase in Ti-Nb has been investigated most thoroughly by BALCERZAK and SASS [BAL71, BAL72]. Some of their results have been presented in *Fig. 4-14*, a series of diffraction patterns which depict a gradual change in the ω reflections from sharp to diffuse with increasing Nb concentration. According to those authors, alloys of compositions between 18 and 25 at.% Nb give rise to sharp ω reflections. This result agrees with that of BAGARIATSKII *et al.* [BAG59], who noted that the threshold of ω formation in quenched alloys was 18 at.% Nb. HICKMAN [HIC69a], as the result of a series of aging experiments, had shown that at 450°C the isothermal ω phase eventually attained compositions of 6 to 11 at.% Nb, which should correspond to the isothermal threshold. In a fairly recent investigation, GUSEVA and DOLINSKAYA [GUS82] noted, as a result of x-ray and other measurements on quenched alloys, that the minimum and maximum critical concentrations for the appearance of athermal ω phase were, respectively, 12 and 20 at.% Nb.

It is especially difficult to make categorical statements about the occurrence of ω phase in higher concentration alloys, to some extent because the quenched product depends not only on the quench rate (as is generally true) but also on the temperature at which the sample is held immediately prior to the quench [BAL72]. Thus, whereas the structure of Ti-Nb(18.4 at.%) was $\omega + \beta$ after oil-quenching from 1000°C, it transformed to α'' when quenched from 900°C. Likewise, Ti-Nb(22.5 at.%) became $\omega + \beta$ during a quench from either 900 or 1000°C and $\alpha'' + \beta$ if the pre-quench temperature was reduced to 800°C. From a heat-content and heat-transfer standpoint, it is reasonable to expect that the actual quench rate experienced by the initially hotter sample should be the slower. Thus, in terms of their effects on the quenched microstructure there may not be very much of a distinction between "quench rate" and "prior annealing temperature". HICKMAN [HIC69a] achieved relatively rapid cooling rates (10^3 °C s^{-1}) during the He-gas quenching of resistively heated strips of alloy. In so doing, he was able to suppress ω phase in all the alloys examined, viz, Ti-Nb($\gtrless 22$

at.%). In particular, Ti-Nb(22 at.%) yielded $\alpha'' + \beta$, in agreement with the results of the 800°C quench of BALCERZAK and SASS [BAL72]; gas-quenched Ti-Nb(25 at.%) was also $\alpha'' + \beta$, whereas the structure of Ti-Nb(25.6 at.%) oil-quenched from 900°C by BALCERZAK and SASS [BAL72] was essentially $\omega + \beta$. HICKMAN's Ti-Nb(27 at.%) was found to be "all β" [HIC69a], while for Ti-Nb(34 at.%) the quenched structure obtained by BALCERZAK and SASS [BAL72], although designated "β", yielded an SAD pattern exhibiting so-called "lines of intensity". These lines, which are also present in association with ω reflections in the lower concentration alloys, persisted in decreasing intensity through Ti-Nb(57 at.%) and have actually been noted in pure Nb [SAS72]. It is permissible to regard the lines of intensity as arising from "diffuse ω" since they are believed to originate from the influence on the bcc lattice of that same vibrational state [SAS72] which gives rise, in the lower concentration alloys, to the athermal ω phase itself.

(b) The Multicomponent β Alloys. The multicomponent β alloys have been discussed in Sect. 3.3.4 with particular reference to their equilibrium properties, which are of course achieved only after aging. The β alloys result from the inclusion in their formulations of sufficient β stabilizer to drop their M_s temperatures to well below room temperature. The properties of quenched multicomponent β alloys are comparable to those of their binary counterparts, e.g., Ti-12Mo in the case of β-III [BOY74]. It has long been known that the presence of solutes such as Al have a retarding influence on the $\beta \to \omega$ reaction [JAF58, p. 141]. Thus, although the presence of Fe and Al in Ti-10V-2Fe-3Al and of Zr and Sn in Ti-11.5Mo-6Zr-4.5Sn (β-III) does not completely prevent the formation of ω phase during quenching from above the $\beta/(\alpha + \beta)$ transus [DUE80a][RAC70], the fact that the ω reflections in the latter alloy are diffuse rather than sharp, as they are in Ti-11.6Mo itself [BAL68], indicates that the simple metals do inhibit the transformation. Since the diffusionless athermal ω transformation is a response to bcc lattice instability, this inhibiting effect by SM additions in general [WIL71] is presumably a result of their increasing the stiffness of the bcc lattice.

(i) Ti-10V-2Fe-3Al. The commercial alloy Ti-10V-2Fe-3Al, (i.e., "Ti-10-2-3"), whether or not it is "near-β" [TOR80] or "β" [DUE80a], is certainly a modified Ti-V alloy just on the threshold of β stability. Its e/a ratio of 4.11_4* identifies it with Ti-V(11 at.%), whose β phase is not fully retained on quenching [MOL65, p. 14][COL75b]. According to TORAN and BIEDERMAN [TOR80], the M_s temperature of Ti-1023 is 555°C, and quenching from above the β transus (788°C) yields both martensite and ellipsoidal ω-phase particles. On the other hand, DUERIG, MIDDLETON,

---

*$(e/a) = 4.00 + (0.0954 \times 1) + (0.0186 \times 4) - (0.0559 \times 1) = 4.11_4$.

TERLINDE, and WILLIAMS [DUE80[a]] claimed that quenching into water, presumably at room temperature, yielded β plus athermal ω phase. If, following DUERIG et al., room temperature lies between M_d (the transus for deformation martensite) and M_s, it is possible that mechanical stress, if it takes place during quenching, could be responsible for a partial martensitic transformation.

In any case, whether quenching stress or deliberately applied post-quench mechanical stress is responsible for a martensitic transformation in this alloy, it is interesting to note that the transformed structure and the ω phase are coexistent. Thus, as DUERIG et al. [DUE80[a]] have noted, the transformation to deformation α'' leaves the ω particles intact; of course, the ω phase is equally durable during mechanical twinning. In other systems it has been noted that martensitic transformation obliterates the ω phase. The energetics of its stability in this case have been discussed by DUERIG et al. [DUE80[a]].

(ii) β-III. The commercial alloy β-III (Ti-11.5Mo-6Zr-4.5Sn) is generally used in the $\alpha + \beta$ aged condition since athermal ω in the quenched alloy, and isothermal ω in an alloy aged in service, are precipitation hardeners which, although they increase the flow stress, reduce the ductility (i.e., they embrittle the alloy). Consequently, there has been little inducement to investigate the fundamental properties of as-quenched β-III. One such study has, however, been undertaken by RACK, KALISH, and FIKE, the results of whose x-ray, TEM, and other observations of quenched and deformation-transformed material have been reported in considerable detail [RAC70]. Quenching of β-III from above the β transus (755°C) resulted in a bcc lattice and a finely dispersed ω-phase precipitate with a mean particle size of $\gtrsim 30$ Å. No trace of martensitic transformation was observed either as a result of quenching into water, or after subsequent quenching into liquid nitrogen, indicating that M_s was below 77 K. Foils being thinned for TEM observation underwent the usual spontaneous transformation which, however, in this case did not obliterate the ω-phase particles. The athermal ω particles could not be distinctly imaged in the electron microscope, making it impossible to specify their shape [RAC70]. Likewise, selected-area electron diffraction performed on the as-quenched alloys revealed the streaked reflections characteristic of diffuse ω. The occurrence of sharp and diffuse, respectively, athermal ω reflections and their implication with respect to ω-phase precipitation has been considered by SASS and colleagues, some of whose results have been summarized in the foregoing discussions. The morphology of β-III's isothermal ω, which has been photographed many times in aged samples (e.g., [BOY74] [WIL82[a]]), is distinctly ellipsoidal and similar to that present in binary Ti-Mo.

5

Mechanical Properties

5.1 Elastic and Plastic Properties of Titanium Alloys at Low and High Temperatures

As indicated in Chapter 1, technical Ti-base alloys fall into three categories: α, $\alpha + \beta$, and β. Unalloyed Ti, α alloys such as Ti-5Al-2.5Sn, and near-α $\alpha + \beta$ alloys such as Ti-6Al-4V and Ti-8Al-1Mo-1V are preferred for service at low temperatures where the β phase could otherwise cause embrittlement. Other $\alpha + \beta$ and β alloys find use in the medium-temperature and "high-temperature" ($\gtrsim 500°C$) ranges. The α phase is stabilized by simple metals such as Al and Sn and interstitial elements such as C, N, and O—in other words, by non-transition elements. The α stabilizers are also rapid solution strengtheners of the alloy in which they are dissolved. The β phase is stabilized by transition elements, which also provide weak solution strengthening. Although the transition elements are much less potent (rapid) strengtheners than the α stabilizers (on a per-atom basis), this deficiency is more than compensated for by their greater solid-solubility range. It therefore turns out that, whereas the strength of the α phase is limited to 80~100 ksi, that of the β phase may be as high as 100~120 ksi [JAF73[a]]. To a first approximation, the strengths of $\alpha + \beta$ alloys are a mixture-rule average of those of their constituents [MAR60, p. 291]. The influence of microstructure on strength is a subject of perennial interest [JAF58, pp. 149 *et seq.*][MAR60, pp. 291 *et seq.*][JAF73[a]] [CHE80]. This is particularly true of $\alpha + \beta$ alloys, for which thermomechanical process variation offers a wide range of microstructural states. On the other hand, apart from the special properties associated with precipitated α_2, the mechanical properties of the α alloys are not so sensitive to microstructure [JAF73[a]].

As pointed out above, the β alloys are unsuited for low-temperature applications; but whereas the α alloys perform satisfactorily at low temperatures, their mechanical properties decrease rapidly with increasing temperature, *Fig. 5-1*. Some $\alpha + \beta$ alloys, such as Ti-6Al-4V and Ti-8Al-1Mo-1V, can also be regarded as having properties suitable for a wide range of cryogenic applications [SAL79].

As the temperature increases above room temperature, since the strengths of the all-α alloys such as Ti-5Al-2.5Sn continue to decrease rapidly, they must be abandoned in the intermediate-temperature range in favor of the $\alpha + \beta$ or β alloys which are capable of maintaining theirs, see *Fig. 5-2*.

With regard to high-temperature service, although all-β technical alloys such as β-III and the more complex β-C are available, general discussions of heat-resistant alloys (e.g., [Pos81]) do not emphasize their use. In fact, the tendency is not only to use an $\alpha + \beta$ alloy at elevated temperatures, but also to select one whose low β content places it in the near-α category.

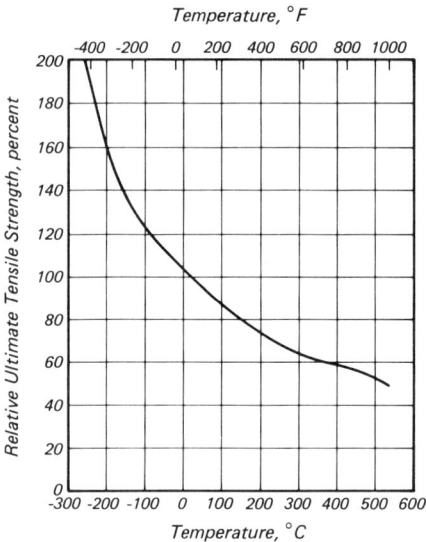

Fig. 5-1. Temperature dependence of the relative ultimate tensile strength of annealed Ti-5Al-2.5Sn (in the form of sheet) [Woo72, p. 5-2:72-4].

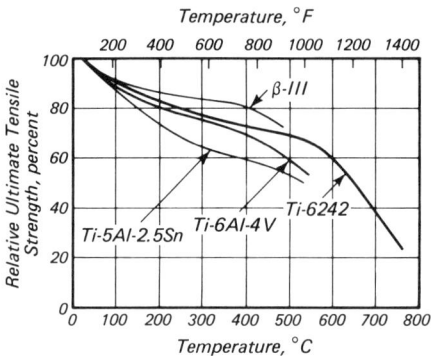

Fig. 5-2. Intercomparison of the temperature dependences of the relative ultimate tensile strengths of four commercial Ti alloys [Woo72].

According to POSTANS and JEAL [Pos81], for example, the alloys best suited to gas-turbine engine use (compressor discs and blades) are Ti-8Al-1V-1Mo (limit 400°C), Ti-4Al-2Sn-4Mo-0.5Si (IMI 550, limit 450°C), Ti-6Al-2Sn-4Zr-2Mo (limit 450°C), Ti-2.5Al-11Sn-5Zr-1Mo-0.2Sn (IMI 679, limit 450°C), Ti-6Al-2Sn-4Zr-2Mo-0.1Si (limit 510°C), and Ti-6Al-5Zr-0.5Mo-0.2Si (IMI 685, limit 520°C).

This chapter discusses the elastic properties (the moduli) and the plastic properties (the strengths) of Ti-base alloys in that order. In so doing, it deals with mechanical properties as measured using: (i) the static techniques of hardness measurement and tensile testing, and (ii) the vibrational or acoustic techniques of dynamic elastic modulus measurement. The static modulus is an engineering number. It is introduced into this chapter in tabular form with no discussion. The results of dynamic modulus measurement lend themselves to discussion in terms of fundamental alloy theory and in this vein are considered in detail in the third section of this chapter. Two important strengthening mechanisms—solution strengthening and precipitation strengthening—are reviewed briefly in Sect. 5.4. Hardness is considered in Sect. 5.5; it is a measurement, simple to perform, whose results are related to both elastic modulus and yield strength. In this sense it couples the results of the preceding two sections. It would be logical to describe next the normal tensile properties of Ti alloys. This again is an engineering topic more suited to a handbook than a general descriptive text such as this. Accordingly, Sect. 5.13, under the heading "Tensile Strengths of Some Commercial Titanium Alloys", confines itself to some yield-strength temperature-dependence data for a few representative technical alloys. The chapter concludes with a brief survey of "anomalous" tensile properties, viz, those which exhibit pronounced departures from stress-strain linearity and/or reversibility.

Part 1: Elastic Properties

5.2 Static Elastic Moduli

The Young's modulus, E, being the slope of the linear portion of the stress-strain, $\sigma(\epsilon)$, curve may be obtained from the results of a static or quasistatic (i.e., very low frequency) tensile test. Alternatively, the slope about the origin of $\sigma(\epsilon)$ may be obtained from a measurement of the velocity of sound in the sample via an appropriate form of the general relationship:

$$\text{sound velocity} = \sqrt{\text{modulus/density}} \qquad (5\text{-}1)$$

"Dynamic moduli" obtained using such approaches will be discussed in the following sections.

Returning to the static moduli, the subject of this section, *Fig. 5-3* serves as a reminder of the definitions of the engineering quantities: bulk modulus (K), shear modulus (G), Young's modulus (E), and the Poisson's ratio (v), which in an isotropically elastic solid (fine, randomly textured grains) are simply related to each other according to:

$$K = \frac{E}{3(1-2v)} \ , \tag{5-2}$$

$$G = \frac{E}{2(1+v)} \ . \tag{5-3}$$

In the literature, mechanical properties are found expressed in one or other of the units: $kg\,mm^{-2}$, $N\,m^{-2}$, $dyne\,cm^{-2}$, Mbar, and GPa. The numerical relationships between them are:

$$1\,N = 10^5\,dyne$$

$$1\,N\,m^{-2} = 1\,Pa$$

$$10^7\,N\,m^{-2} = 1.0197\,kg\,mm^{-2}$$

$$10^{11}\,N\,m^{-2} = 10^{12}\,dyne\,cm^{-2}$$

$$= 100\,GPa$$

$$= 1\,Mbar$$

Table 5-1 is a listing of the Young's moduli of several technical Ti-base alloys.

5.3 Dynamic Elastic Moduli: Long-Wavelength Methods

The wavelengths of kHz-frequency vibrations propagated along a metallic bar or wire are commensurate with its length. In either its longitudinal (L) or transverse (T) vibrational modes, since unconstrained stretching and relaxation is taking place in either case, the wave velocity, v, along the sample is controlled simply by the "static" Young's modulus, E, according to $v_{L,T}$ (long wavelength) $= \sqrt{E/\rho_d}$, where ρ_d is the density. Torsional

Bulk Modulus, K

$$K = \frac{\sigma_{hyd.}}{\epsilon_{vol.}}$$

Shear Modulus, G

$$G = \frac{\tau}{\gamma}$$

Young's Modulus, E

$$E = \frac{\sigma}{\epsilon}$$

$\gamma = \tan\theta$

$\epsilon/2$

Poisson's Ratio, $\nu = \dfrac{-\epsilon_y}{\epsilon_z}$

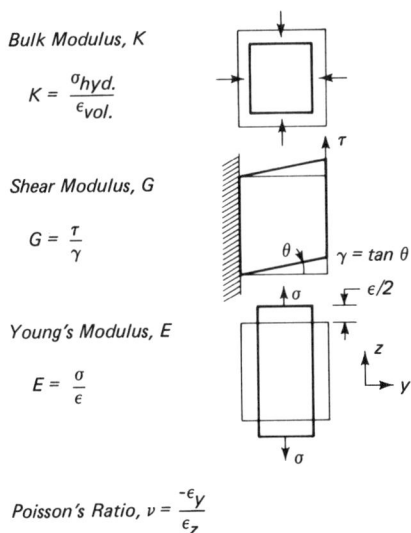

Fig. 5-3. Standard definitions of the elastic moduli of polycrystalline solids.

vibrations are required for the determination of the shear modulus according to $\nu_{torsion} = \sqrt{G/\rho_d}$. A commercially available device often used for this type of measurement is the "Elastomat" designed by F. FORSTER (see [FED63]). Alternatively, the Marx-oscillator technique may be employed. In recent measurements of Ti-Nb samples using the latter method [LED81], cylindrical-rod specimens, about 5 mm in diameter, were cemented to a matched pair of Y-cut, X-plated, rectangular-rod 50 kHz quartz crystals. Using suitable electronics, this three-component composite oscillator (the Marx oscillator) was swept in frequency until the half-wave resonance took place. This was detected by an oscilloscope and measured by a frequency meter. The Young's modulus could then be determined from the relationship:

$$E = 4\rho_d f^2 l^2 \ , \tag{5-4}$$

in which f is the resonant frequency of the rod of length l.

5.4 Systematic Variation of Elastic Moduli with Composition and Microstructure in Ti-Base Alloys

Using the Elastomat method, FEDOTOV and colleagues have measured the Young's and shear moduli of numerous series of α-stabilized Ti-SM and β-stabilized Ti-TM binary alloys as functions of composition, hence

Table 5-1. Elastic Moduli of Several Commercial Titanium-Base Alloys: Typical Room-Temperature Values [STR82]

Alloy name	Nominal composition	Condition	Young's modulus, E Tensile 10^6 psi	10^{10} N·m^{-2}	Compressive 10^6 psi	10^{10} N·m^{-2}	Shear modulus, G 10^6 psi	10^{10} N·m^{-2}
5-2.5	Ti-5Al-2.5Sn	Annealed ($\frac{1}{4}$-4h)/(1300-1600°F)	15.8	10.9	—	—	7.0	4.8
3-2.5	Ti-3Al-2.5V	Annealed (1-3h)/(1200-1400°F)	14.5	10.0	15.0	10.3	—	—
6-2-1-1	Ti-6Al-2Nb-1Ta-1Mo	Annealed ($\frac{1}{4}$-2h)/(1300-1700°F)	16.8	11.6	18.0	12.4	3.5	2.4
8-1-1	Ti-8Al-1Mo-1V	Annealed 8h/1450°F	17.5	12.1	18.0	12.4	6.7	4.6
Corona 5	Ti-4.5Al-5Mo-1.5Cr	α-β annealed after β processing	15.5-17.0	10.7-11.7	—	—	—	—
Ti-17	Ti-5Al-2Sn-2Zr-4Mo-4Cr	α-β or β processed plus aged	16.3	11.2	16.1	11.1	6.1	4.2
6-4	Ti-6Al-4V	Annealed 2h/(1300-1600°F)	16.0	11.0	16.6	11.4	6.1	4.2
		Aged	16.5	11.4	—	—	6.1	4.2
6-6-2	Ti-6Al-6V-2Sn	Annealed 3h/(1300-1500°F)	16.0	11.0	—	—	6.5	4.5
		Aged	17.0	11.7	17.5	12.1	6.5	4.5
6-2-4-2	Ti-6Al-2Sn-4Zr-2Mo	Annealed 4h/(1300-1550°F)	16.5	11.4	18.0	12.4	—	—
6-2-4-6	Ti-6Al-2Sn-4Zr-6Mo	Annealed 2h/(1500-1600°F)	16.5	11.4	18.0	12.4	—	—
6-22-22	Ti-6Al-2Sn-2Zr-2Mo-2Cr-0.25Si	α-β processed plus aged	15.7	10.8	16.1	11.1	6.7	4.6
10-2-3	Ti-10V-2Fe-3Al	Aged	15.9	11.0	16.3	11.2	—	—
15-3-3-3	Ti-15V-3Cr-3Sn-3Al	Aged	14.3	9.9	15.9	11.0	6.2	4.3
13-11-3	Ti-13V-11Cr-3Al	Annealed $\frac{1}{2}$h/(1400-1500°F)	14.3	9.9	15.2	10.5	—	—
		Aged	16.0	11.0	15.8	10.9	—	—
38-6-44	Ti-3Al-8V-6Cr-4Mo-4Zr	Annealed $\frac{1}{2}$h/(1500-1700°F)	12.5	8.6	—	—	—	—
		Aged	16.7	11.5	15.0	10.3	5.8	4.0
β-III	Ti-4.5Sn-6Zr-11.5Mo	Annealed $\frac{1}{2}$h/(1300-1600°F)	12.0	8.3	11.0	7.6	3.9	2.7
		Aged	15.0	10.3	16.0	11.0	5.9	4.1

of composition-related microstructure. A representative group of results is presented and briefly discussed in the following two sections.

5.4.1 The Elastic Moduli of Ti-Al Alloys: Long-Wavelength Results

The Young's modulus, E, and the shear modulus, G, have been measured by FEDOTOV [FED66] on a series of Ti-Al alloys whose composition range, 0-40 wt.% Al (i.e., 0-54 at.%), includes both the α_2 phase (Ti$_3$Al) and γ phase (TiAl) intermetallic compounds. Since $E \cong 2.6\ G$ (according to Eqn. (5-2), assuming $\nu \cong 0.3 = $ const.), it is to be expected that the curves of E and G versus some common parameter would be "parallel". This indeed turns out to be the case, and several examples of it are presented below. Only one of the moduli, E, is shown in *Fig. 5-4*, which emphasizes two important properties of the Ti-Al system: (i) the addition of Al is responsible for a rapid increase in modulus; and (ii) anomalies appear at compositions corresponding to Ti$_3$Al (a local maximum) and TiAl (a point of inflexion). Similar anomalies have also appeared in the plot of θ_D versus Al concentration, *Fig. 2-14*. This modulus-θ_D parallelism is to be expected in view of the fact that values of θ_D can be calculated from the results of sound-velocity measurements in the manner to be discussed in Sect. 5.6.3. As suggested in Sect. 2.4.1, the actual or incipient maxima in E and θ_D are indicative of the lattice stiffening that occurs when the natural tendency for bond directionality characteristic of Ti-SM alloys "sharpens up" in the vicinity of the stoichiometric compositions [COL82^a].

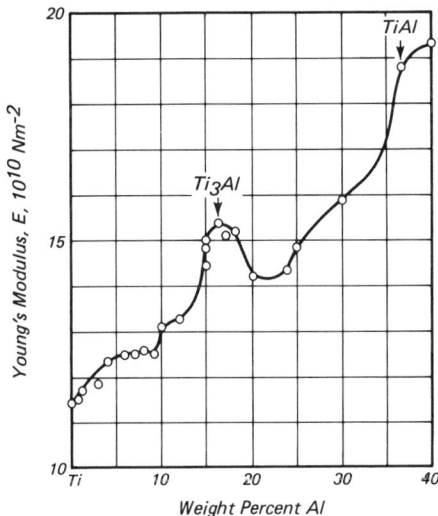

Fig. 5-4. Young's modulus, E, of Ti-Al as a function of Al concentration [FED66, p. 208].

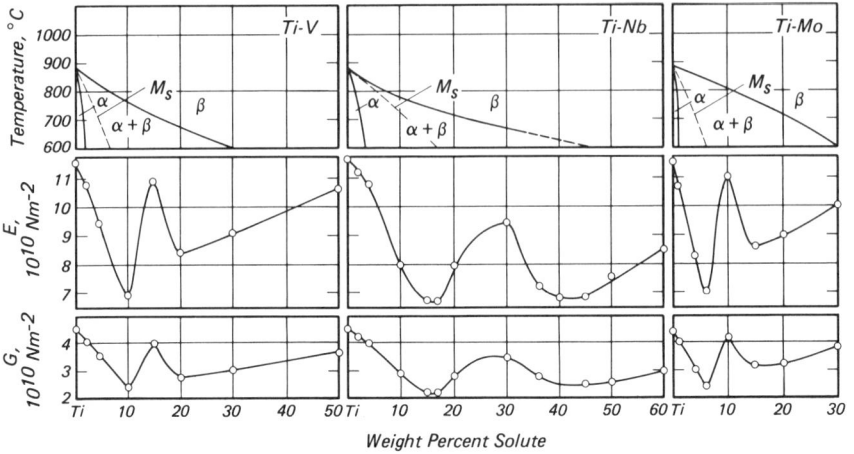

Fig. 5-5. Young's modulus, E, and shear modulus, G, as a function of composition-related microstructure in Ti-V, Ti-Nb, and Ti-Mo alloys [FED73].

5.4.2 The Elastic Moduli of Ti-TM Alloys: Long-Wavelength Results

(a) The β-Isomorphous Alloys: Ti-V, Ti-Nb, and Ti-Mo. Using the long-wavelength transverse, longitudinal, and torsional resonances of alloy rods, FEDOTOV and colleagues [FED63, FED64, FED66, FED73] have measured as functions of composition the Young's moduli, E, and the shear moduli, G, of the β-isomorphous alloys—Ti-V, Ti-Nb, and Ti-Mo—in both the quenched (from 24h/900°C) and quenched-plus-annealed (200h/700°C plus 500h/600°C) conditions. The results for the quenched alloys are depicted in *Fig. 5-5*, where they can be compared with the accompanying equilibrium/nonequilibrium phase diagrams. Immediately obvious is the expected parallelism of the composition dependences of E and G (see Sect. 5.6.3). Further discussions of the figures must be in terms of the nonequilibrium phases α^m and athermal ω which form on quenching.

Attention is first of all drawn to the Ti-rich alloys whose structures are first α' and then α'' with increasing solute concentration.* In the martensitic regime, each alloy exhibits rapid softening as the solute concentration increases. Assuming the correctness of the data of *Table 4-4** (for the composition of the α'/α'' phase boundary), it seems that whereas the change from the α' to the α'' structure does not interrupt the E_{α^m} versus composition curve for Ti-Mo and Ti-Nb, the same is not true for the Ti-V

---

*The compositions of the α'/α'' boundaries are given in *Table 4-4*. There is some disagreement about the existence of a quenched-α'' variant in Ti-V [WIL73][FLO82] (see Sect. 4.2.3).

system. Of course it would be useful to know to what extent the possible existence of mixed phases — and in particular the presence of ω phase — influences the upturn shown by each of the curves.

Transferring attention to the β-phase alloys — on the right-hand side of each diagram — as solute content is *decreased* the β phase becomes continuously softer until the product of that instability, ω-phase precipitation, eventually makes its presence felt by stiffening the lattice. According to BAGARIATSKII et al. [BAG59], *Table 4-2(a)*, the solute concentrations corresponding to which (athermal) ω phase is formed on quenching are: Ti-V, 13 at.% (14 wt.%); Ti-Nb, 18 at.% (30 wt.%); Ti-Mo, 4.5 at.% (8.6 wt.%). These values are in excellent agreement with the positions of the E- and G-modulus peaks.

Numerous other physical properties respond in a like manner to the composition-related microstructures. Since the Debye temperature can be synthesized from the macroscopic elastic moduli (see below), a parallelism between θ_D and E or G is expected and, according to *Fig. 5-6*, is indeed observed. The behavior of the Vickers-hardness curve also exemplifies the

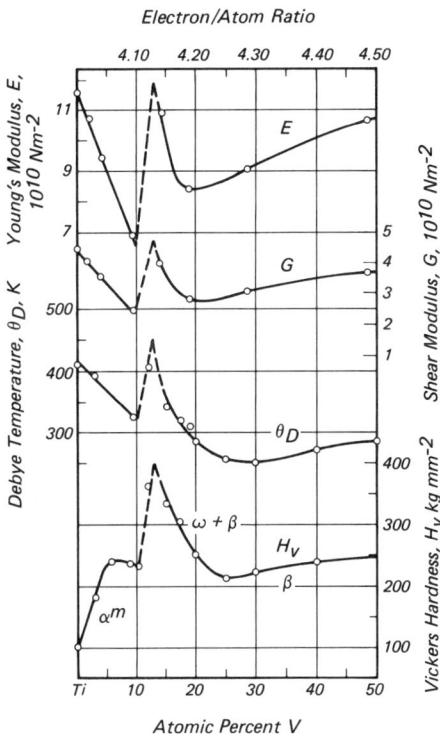

Fig. 5-6. Young's modulus, E, shear modulus, G, Debye temperature, θ_D, and Vickers hardness, H_V, as functions of composition in quenched Ti-V alloys. *References*: [FED73] (E and G), [COL75^c] (θ_D), and [COL84] (5-kg diamond-pyramid hardness, H_V).

connection (to be considered below) among hardness, strength, and modulus. As solute content decreases in the bcc field, the composition dependences of the four parameters plotted in *Fig. 5-6* respond to the stiffening and hardening influences of ω-phase precipitation. On the α''' side, only the hardness data, particularly at low solute concentrations, exhibit departures from parallelism in a manifestation of some kind of competition between solution strengthening and lattice softening.

The influence of heat treatment on modulus is considered in *Fig. 5-7* with reference to the equilibrium and metastable-equilibrium phase diagrams of Ti-Nb, constructed from data sources referred to in the caption. The form of the E versus composition curve for the 24h/900°C/WQ alloys has already been discussed. After the alloy series has been annealed according to the prescription 200h/700°C + 500h/600°C, the modulus data give rise to a new curve designated *A-B-C-D*. The segment *A-B*, which represents equilibrium-α phase, naturally follows the "as-quenched" data; the segment *C-D*, which is in the 600°C equilibrium-β field, also follows the old data; while *B-C*, for the equilibrium-$\alpha+\beta$ field, is simply a "tie-line". The temperature-time results are also interesting: with a metastable alloy, depending on the decomposition kinetics, a temperature-dependence experiment may also be a short-time aging experiment. For

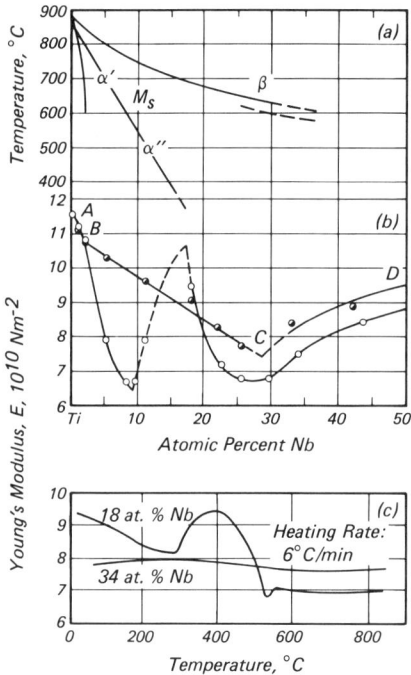

Fig. 5-7. Young's moduli, E, of quenched and aged Ti-Nb alloys as a function of metallurgical condition. *(a)* Equilibrium transi from standard sources (see *Fig. 3-10*) and an M_s line from [JEP70]. *(b)* Young's moduli of quenched (from 24h/900°C) and quench-plus-aged (100h/800°C + 200h/700°C + 500h/600°C) Ti-Nb alloys [FED73] (see also [FED64]). *(c)* Change of modulus in response to heating at the rate of about 6°C per min [FED64].

example, quenched Ti-Nb(18 at.%) possesses a high volume-fraction of ω phase (*Table 4-2*); upon heating through 300°C, solute diffusion becomes active and additional isothermal precipitation takes place, resulting in a modulus peak centered about 400°C, an optimal temperature for isothermal ω-phase precipitation. At higher temperatures some α precipitation commences, enriching the matrix with Nb and lowering E, although within the 1 h which elapses as the temperature is raised from 400 to 800°C, thermodynamic equilibrium is not achieved. The experiment illustrates the relatively rapid kinetics of the ω-phase reaction. The equilibrium state of the other alloy represented in the figure, Ti-Nb(34 at.%), is $\alpha + \beta$ at room temperature. But since the reaction kinetics are so sluggish,* the β phase is retained on quenching and the experiment on initially quenched Ti-Nb(34 at.%) measures the actual E-modulus temperature dependence of the quenched β phase. This turns out to be relatively small.

The responses of the mechanical properties of three representative β-isomorphous Ti-TM alloys (viz, Ti-10V, Ti-17Nb, and Ti-6Mo) to quenching from various temperatures within the interval 600-900°C have been investigated by JAMES and MOON [JAM 70]. Hardness measurements and tensile tests were conducted, Young's modulus was determined using an Elastomat-type of instrument similar to that employed by FEDOTOV et al. (see above), while the internal friction (i.e., the imaginary component of the complex elastic modulus) was calculated from the decay time constant of the resonant vibrations of rods. The quantities listed generally turned out to be strongly dependent on the pre-quench temperature: In Ti-12V, for example, as that temperature was dropped from 895°C to 659°C, although the hardness remained fairly constant at about 200 kg mm^{-2} down to 709°C (pre-quench), it rose steeply to 405 kg mm^{-2} at 700°C, then went on to decrease monotonically with further reduction in the pre-quench temperature. The results were interpreted in the following way: (i) In response to quenching from temperatures above the $(\alpha + \beta)/\beta$ transus, the quenched product is α' martensite plus retained interplatelet β; the resulting low strength and modulus are attributable to stress-induced martensitic transformation of the metastable-β phase. (ii) When the pre-quench temperature is dropped below $(\alpha + \beta)/\beta$, ω-phase precipitation in the β component is responsible for the observed pronounced increases in hardness, yield strength, and modulus. (iii) The decrease in these quantities with further drop in pre-quench temperature corresponds to the establishment of an equilibrium-β component too rich to support any ω-phase precipitation.

---

*Hence the need for the 500 h equilibration time at 600°C referred to above.

(b) The β-Eutectoid Alloys: Ti-Cr, Ti-Mn, Ti-Fe, Ti-Co, and Ti-Ni. The elastic properties of the β-eutectoid alloys of Ti with first-row transition elements have been studied by FEDOTOV and colleagues [FED73]. The modulus-composition dependences of alloys quenched from 1000°C are intercompared in *Fig. 5-8*. As before, with decreasing solute content, the alloys all show an elastic softening. This is followed by a rapid increase in stiffness in the composition range where athermal ω phase is expected. In the martensitic regime, the Young's modulus composition dependences of the β-eutectoid alloys are generally much smaller than those exhibited by the β-isomorphous series. Secondly, a trend manifests itself within the eutectoid series. Particularly for the Ti-Fe, Ti-Co, and Ti-Ni trio, the relatively flat composition dependences are believed to be due to departures from the single-phase martensitic structure as a consequence of partial decomposition of either (i) the β phase above M_s or (ii) the martensitic phase below it during the quenching process, as a result of the very high diffusivities in Ti of Fe, Co, and Ni. The special positions occupied by these three elements, with regard to diffusivity, have already been discussed in Sect. 2.4.4 (see *Fig. 2-18*).

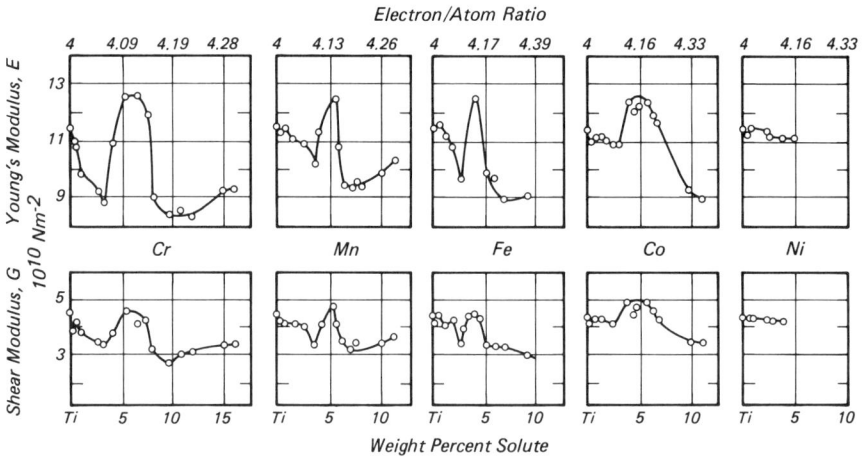

Fig. 5-8. Young's moduli, E, and shear moduli, G, of β-quenched alloys of Ti with the "β-eutectoid-forming" solutes Cr, Mn, Fe, Co, and Ni [FED73].

5.5 Dynamic Elastic Modulus: Ultrasonic Methods

5.5.1 Basic Theory

Using ultrasonic techniques in the 20-100 MHz frequency range it is possible to generate sound waves of wavelength $\sim 10^{-4}$ m, very much smaller than the physical dimensions of the usual specimen. By applying Eqn.

(5-1) in the form $v_{ij} = \sqrt{C_{ij}/\rho_d}$ to a series of appropriately cut monocrystalline samples it is, in principle, possible to separately evaluate the individual components of the 6×6 stiffness matrix* $[C_{ij}]$ defined by:

$$[\sigma] = [C_{ij}][\epsilon] \; , \tag{5-5}$$

where σ and ϵ have their usual meanings of stress and strain, respectively. Although the ultrasonic methods may be replaced by low-frequency techniques in the measurement of the elastic moduli of polycrystalline materials, their use is almost mandatory if the elastic moduli of single crystals are required. In general linear elastic theory, the $[C_{ij}]$ or $[S_{ij}]$ may be represented in Voigt's contracted notation by the 6×6 matrix:

$$\begin{bmatrix} 11 & 12 & 13 & 14 & 15 & 16 \\ 21 & 22 & 23 & 24 & 25 & 26 \\ 31 & 32 & 33 & 34 & 35 & 36 \\ 41 & 42 & 43 & 44 & 45 & 46 \\ 51 & 52 & 53 & 54 & 55 & 56 \\ 61 & 62 & 63 & 64 & 65 & 66 \end{bmatrix} \tag{5-6}$$

For a crystal of cubic symmetry, this reduces to:

$$\begin{bmatrix} 11 & 12 & 12 & & & \\ 12 & 11 & 12 & & \bigcirc & \\ 12 & 12 & 11 & & & \\ & & & 44 & & \\ & \bigcirc & & & 44 & \\ & & & & & 44 \end{bmatrix} \tag{5-7}$$

leaving only three independent monocrystalline elastic moduli. Less symmetric crystals have up to 21 independent elastic moduli.

---

*Or a compliance matrix, $[S_{ij}]$, may be derived from Eqn. (5-5) by matrix inversion: $[\epsilon] = [S_{ij}][\sigma]$.

5.5.2 The Stiffness Moduli of Cubic Monocrystals

As shown in matrix (5-7), a cubic monocrystal is characterized by three fundamental stiffness moduli: C_{11}, C_{44}, and C_{12}. In practice, the first two are measured directly, while C_{12} is obtained in association with the others. An important shear stiffness modulus is $C' = (C_{11} - C_{12})/2$ which, although it is made up of two of the fundamental moduli, is actually obtainable directly by experiment. The propagation descriptions of the ultrasonic waves needed for the measurement of C_{11}, C_{44}, and C' are:

(a) C_{11}: a longitudinal wave in a $\langle 100 \rangle$ direction.

(b) C_{44}: a transverse wave in a $\langle 100 \rangle$ direction polarized along $\langle 010 \rangle$, or a transverse wave in a $\langle 110 \rangle$ direction polarized along $\langle 001 \rangle$.

(c) C': The other transverse wave in a $\langle 110 \rangle$ direction, polarized along $\langle \bar{1}10 \rangle$.

5.5.3 The Significance of C' for Transition-Metal Alloys

Since C_{44} is governed by the transverse $\langle 110 \rangle$ wave, $\langle 001 \rangle$-polarized, and C' by the same wave, $\langle \bar{1}10 \rangle$-polarized, in an isotropic solid $C_{44} = C'$ (see also [Fis75, p. 202]). Departure from isotropy is gauged by the value of the Zener "anisotropy ratio", $A \equiv C_{44}/C'$. In bcc simple metals, A can be quite large, for example $A_{Na} = 7.5$, whereas for the bcc transition elements of the sixth group, Cr, Mo, and W, the anisotropy ratio takes on the values 0.71, 0.72, and 1.01, respectively [Fis75], indicating the influence of the d-band electrons in stabilizing the bcc lattice. Fisher has discussed in detail the physical significances of A and C' in relationship to electronic factors which govern bcc stability [Fis70, Fis75][Kat79]. He has pointed out that, whereas the C_{44} shears are resisted primarily by nearest-neighbor repulsion, the C' shear stiffness depends primarily on the next-nearest-neighbor forces. This explains its low values in simple metals (e.g., 0.0065×10^{11} N m^{-2} for Na) [Fis75]. On the other hand, additional cohesive contributions from d-electrons are thought to be responsible in transition elements for the observed large values of C'. The C' modulus appears to be interpretable as a bcc stability parameter; thus, for the highly stable bcc transition elements Cr, Mo, and W, $C' \cong 1.5 \times 10^{11}$ N m^{-2}, but with decreasing electron/atom ratio C' decreases rapidly, tending to zero at room temperature for alloys which possess ω-phase instabilities or transform martensitically to hcp at ordinary temperature [Col73a]. This continuously decreasing bcc lattice stability with decreasing e/a is illustrated in *Fig. 3-3* with data for the pure elements Cr, Mo, and W (group VI); V, Nb, and Ta (group V); and members of the alloy systems Zr-Nb, Ti-Cr, and Mo-Re. It is interesting to note that C' seems to be a universal function of e/a [Fis70].

5.6 Ultrasonic Measurements of the Macroscopic Elastic Moduli

5.6.1 The Isotropic Solid

All propagation directions being equivalent in the isotropic solid, the only variables are those embodied in the vibrational modes themselves. These, which may be either transverse or longitudinal, call for only *two* elastic moduli, best represented by the Lamé parameters λ and μ [JAE62, p. 54]. The modulus for longitudinal-wave propagation is $\lambda + 2\mu$, that for transverse-wave propagation is μ. The macroscopic moduli are expressible in terms of the Lamé parameters in the following way:

$$K = \lambda + \frac{2}{3}\mu \ , \tag{5-8}$$

$$G = \mu \ , \tag{5-9}$$

$$E = \frac{9KG}{3K + G} = \frac{\mu(3\lambda + 2\mu)}{\lambda + \mu} \ , \tag{5-10}$$

$$\nu = \frac{E}{2G} - 1 = \frac{\lambda}{2(\lambda + \mu)} \ . \tag{5-11}$$

5.6.2 An Aggregation of Cubic Monocrystals: The VRH Approximation

VOIGT has shown how, starting with the monocrystalline elastic stiffness moduli, C_{ij}, it is possible to derive expressions for the macroscopic shear modulus, G_V, and the bulk modulus, K_V. For a macroscopically isotropic aggregate of cubic monocrystals, the VOIGT approach yields:

$$5G_V = C_{11} - C_{12} + 3C_{44} \ , \tag{5-12}$$

and

$$3K_V = C_{11} + 2C_{12} \ . \tag{5-13}$$

In a parallel analysis, REUSS has expressed the macroscopic moduli, now G_R and K_R, respectively, in terms of the monocrystalline compliance moduli, S_{ij}. Thus, under conditions similar to the above:

$$\frac{5}{G_R} = 4S_{11} - 4S_{12} + 3S_{44} \ , \tag{5-14}$$

and

$$\frac{1}{3K_R} = S_{11} + 2S_{12} \ . \tag{5-15}$$

Next, through the application of some identities connecting S_{ij} and C_{ij} in cubic crystals, these relationships reduce still further to:

$$5G_V = 2C' + 3C_{44} \ , \tag{5-16}$$

$$\frac{5}{G_R} = \frac{2}{C'} + \frac{3}{C_{44}} \ , \tag{5-17}$$

and

$$K_V = K_R = K = \frac{1}{3} (C_{11} + 2C_{12}) \ . \tag{5-18}$$

HILL showed that the VOIGT and REUSS approximations were, respectively, greater than and less than the true polycrystalline moduli, which were then better represented, in what later became known as the VRH approximation, by the arithmetic means of these extremes.* Thus, as summarized by ANDERSON [AND63]:

$$\langle G \rangle = G_H = (G_V + G_R)/2 \tag{5-19}$$

and

$$\langle K \rangle = K_H = (K_V + K_R)/2 \ , \tag{5-20}$$

in which the HILL arithmetic-mean approach has been adopted. After this, with the aid of Eqns. (5-16) through (5-18), E_H and ν_H can be calculated by means of Eqns. (5-10) and (5-11), and a Debye temperature can be obtained in the manner to be outlined below.

5.6.3 An Aggregation of Cubic Monocrystals: The VRHG Approximation for θ_D

The Debye temperature, θ_D, obtainable experimentally from the lattice contribution to the low-temperature specific heat, may be regarded as the

---

*References to bounds more sophisticated than VRH are given by LEDBETTER [LED80].

ultimate embodiment of all the elastic constants. ANDERSON [AND63] has described how a reliable value of θ_D for an elastically isotropic solid can be calculated from the monocrystalline moduli, using computed values of the macroscopic moduli as an intermediate step.

The expression for θ_D which he recommended is:

$$\theta_D = \frac{h}{k_B} \left(\frac{3\mathfrak{N}}{4\pi} \right)^{1/3} \left(\frac{\rho_d}{M} \right)^{1/3} v_m \quad (\text{K}) \; , \tag{5-21}$$

in which the product of the first two factors (consisting only of atomic constants) is 2.514×10^{-3} cgs units, ρ_d is the density (g cm^{-3}), and M is the mean atomic weight of the alloy. v_m (cm s^{-1}) is an average sound velocity; it may be evaluated using:*

$$v_m = \left\{ \frac{1}{3} \left(\frac{2}{v_T^3} + \frac{1}{v_L^3} \right) \right\}^{-1/3} \; , \tag{5-22}$$

after the "longitudinal" and "transverse" velocities v_L and v_T, respectively, have been deduced from:

$$v_L = \sqrt{(K + (4/3)G)/\rho_d} \; , \tag{5-23}$$

and

$$v_T = \sqrt{G/\rho_d} \; , \tag{5-24}$$

K and G having been calculated from the monocrystalline elastic moduli using the procedures of the previous subsection. The title VRHG was applied by ANDERSON [AND63] to this VRH-based method of calculating θ_D, in recognition of the contribution that GILVARRY had made to the subject.

5.7 The Monocrystalline Elastic Moduli of β-Titanium-Transition-Metal Alloys

Using techniques such as the pulse-superposition method and the pulse-echo-overlap method [KAT79], FISHER and colleagues have measured the

---

*LEDBETTER [LED80] has recommended that attempts be made to replace Eqn. (5-22) with a more elaborate formula in the interests of obtaining more accurate values of θ_D, especially in cases of large elastic anisotropy.

ultrasonic elastic moduli of Ti-V [KAT79, KAT79[a]], Ti-Cr [FIS70, FIS70[a]], and Ti-Nb [REI73]. The results of these measurements, together with those of subsequent VRH calculations of E, are presented in *Table 5-2*.

Finally, as a test of the applicability of the VRHG method, the Debye temperatures of a series of Ti-Cr alloys were calculated from the monocrystalline elastic moduli. The experimentally obtained moduli themselves are plotted as functions of Cr composition in *Fig. 5-9(a)*; the VRGH-calculated θ_D's are listed in *Table 5-2* and plotted in part *(b)* of *Fig. 5-9*. Part *(c)* of that figure displays the calculated θ_D's alongside the experimentally obtained and empirically extrapolated θ_D's of a series of Ti-Mo alloys [COL72[a]]. Taken together, the figures intercompare for each alloy

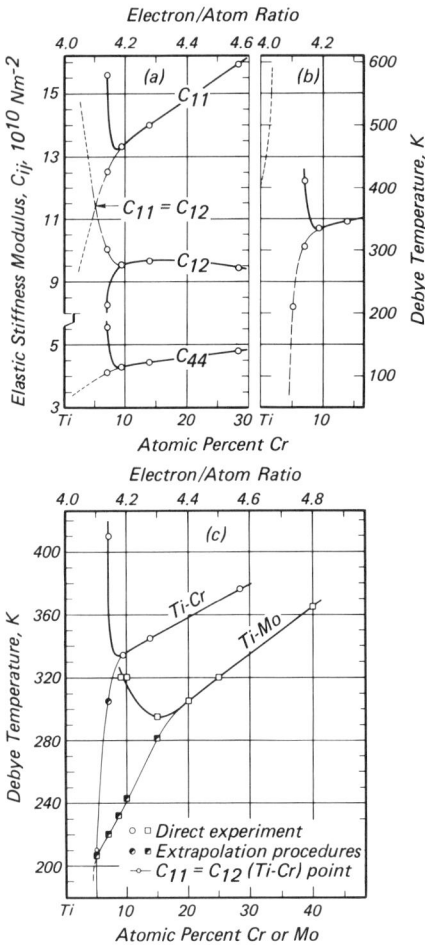

Fig. 5-9. *(a)* Elastic constants of Ti-Cr [FIS70] extrapolated smoothly to the 5 at.% Cr point (dashed lines) and thereafter linearly (chain lines). *(b)* VRHG-calculated θ_D's based on the actual and extrapolated (dashed and chain lines) elastic constants. *(c)* VRHG-calculated θ_D's for Ti-Cr compared with the calorimetrically measured and extrapolated θ_D's for Ti-Mo; see [COL72] for details.

Table 5-2. Monocrystalline Elastic Stiffness Moduli (Elastic Constants) of Ti-V, Ti-Cr, and Ti-Nb

Crystal	Electron/ atom ratio	Condition	Elastic constants,* 10^{10} N m^{-2} (10^{11} dyne cm^{-2})			VRH-calculated Young's modulus,** E, 10^{10} N m^{-2}	Molar weight	VRHG-calculated Debye temperature† Density ρ_d, g cm^{-3}	θ_D, K
			C_{11}	C_{44}	C'				
Ti-V(29.4 at.%)	4.29		14.002	3.966	2.026	8.34	48.79	4.929	328
Ti-V(38.5 at.%)	4.39		14.896	4.095	2.421	9.09	49.07	5.044	340
Ti-V(53 at.%)	4.53		16.760	4.129	3.125	10.09	49.51	5.328	355
Ti-V(73 at.%)	4.73		19.227	4.148	4.060	11.22	50.12	5.640	369
Ti-Cr(6.98 at.%)	4.14	Extrapolated from >900°C to 20°C	12.50	4.10	1.24	7.08	48.19	4.677	305
Ti-Cr(6.98 at.%)	4.14	Brine quench	15.59	5.54	3.67	12.29	48.19	4.677	409
Ti-Cr(9.36 at.%)	4.19	Brine quench	13.31	4.27	1.90	8.45	48.28	4.725	334
Ti-Cr(13.81 at.%)	4.28	Brine quench	13.99	4.42	2.18	9.08	48.47	4.834	345
Ti-Cr(28.37 at.%)	4.57	Brine quench	15.91	4.77	3.25	10.98	49.06	5.027	377
Ti-Nb(40.4 at.%)	4.40	Annealed 65h/1700°C	15.65	3.963	2.247	8.74	66.08	6.189	291
Nb	5.00		24.74	2.80	5.69	10.45	92.91	8.578	268

*Ti-V data, KATAHARA et al. [KAT79]; Ti-Cr data, FISHER and DEVER [FIS70, FIS70a]; Ti-Nb data, REID et al. [REI73]; Nb data, FISHER et al. [FIS75a].

**Calculated from Eqn. (5-10) with ⟨G⟩ and ⟨K⟩ given by Eqns. (5-19) and (5-20).

†Calculated from Eqn. (5-21) with the assistance in turn of Eqns. (5-22), (5-23), (5-24), (5-19), and (5-20). In this c.g.s. representation, the C_{ij} are in dyne cm^{-2}.

In both ** and †, $K_V = K_R$ are given by Eqn. (5-18) and G_V and G_R are given by Eqns. (5-16) and (5-17).

system the measured properties of the actual quenched alloys with those predicted, using certain extrapolation procedures, for what has been referred to as "virtual-β" alloys (i.e., bcc alloys outside their compositional ranges of stability). The extrapolation procedures have been described elsewhere: that for Ti-Cr in [Fis70[a]] and for Ti-Mo in [Col72[a], Col73]. The figure demonstrates that C' vanishes, and θ_D drops to about 200 K (very low for a transition element) at a composition very close to that at which martensitic transformation takes place at room temperature in these systems. The ω-phase precipitate (a product of the precursor instability), which serves to stabilize the lattice, is responsible for the stiffening noted at low values of e/a in C_{11}, C_{44}, and θ_D.

Part 2: Normal Plastic Properties

5.8 Strengthening of Titanium Alloys

Although the principles of solution strengthening must be taken into full consideration in the design of Ti alloys—especially α alloys—the strengths of the heat-treatable $\alpha + \beta$ and β alloys tend to be dominated by "micro-structural effects". The different classes of microstructure, and the associated mechanical properties obtainable as a result of heat treating a single $\alpha + \beta$ alloy, are so numerous that strengthening in such systems, for this reason alone, eludes fundamental analysis. Furthermore, since little information is available on the extent to which alloying elements are distributed between the two phases in such alloys, it is generally not possible to apply solution-strengthening principles to them [Jaf73[a], p. 1680]. As a consequence of these two difficulties, strengthening in heat-treatable alloys can be systematized only from a phenomenological standpoint.

5.8.1 Solid-Solution Strengthening

The two important classes of solution strengtheners are the interstitial elements B, C, N, and O, and simple metals such as Al, Ga, and Sn. Interstitial-element strengthening of Ti and other metals has been considered extensively by Conrad and colleagues (see below). The substitutional α stabilizer/strengtheners, Sn, Ga, and especially Al, either singly or in combination, were the solutes considered by Collings and Gegel in their papers dealing with the electronic and thermodynamic aspects of solution strengthening [Geg73[a]][Col73[a], Col73[b], Col75[a], Col82[a]]. Transition elements cannot be regarded as true solution strengtheners of Ti; naturally they must contribute some measure of strengthening, but that is not their primary role.

(a) Solution Strengthening by α Stabilizers. Solution strengthening by simple metals when dissolved in Ti can be qualitatively understood in terms of the formation of strong, local, and directional electronic bonds between the solute and the surrounding Ti atoms. A moving dislocation experiences a strong pinning force when a segment of its core becomes identified with the local environment of the SM solute atom. Guided by the responses to alloying of physical (electronic) properties such as electrical resistivity, Hall coefficient, and magnetic susceptibility, COLLINGS and co-workers have attempted a qualitative explanation of the nature of the local solute-solvent interaction. As a result, the following description has emerged: In the vicinity of isolated TM atoms dissolved in simple metals, the amplitudes of the d-wavefunctions are large; these are referred to as virtual bound states. This picture carries over to the converse situation, viz, isolated SM atoms dissolved in transition metals (in particular Ti) — with the result that we again find maximum d-wavefunction amplitudes on the transition-metal sites and a tendency for the d-electrons to be excluded from the SM-atom positions. This avoidance of the SM atoms by the d-wavefunctions, a phenomenon which has been demonstrated experimentally in the case of V-Al by VAN OSTENBURG et al. [VAN64], has several consequences: (i) It leads to a representation of the alloy as a "diluted" or "expanded" transition metal, in this case Ti. This in turn, it can be argued, justifies the α-stabilizing property of such solute atoms. (ii) It suggests that the solute atoms should be strong scatterers of conduction electrons, i.e., should be associated with high specific (per atom) solute resistivities. This has indeed been observed. The results of STERN's tight-binding theory of disordered alloys [STE68], augmented by VAN OSTEN-BURG's experimental results for a single member of that class, suggest that in Ti-SM alloys, the s,p valence electrons of the SM atoms contribute to a low-lying band. The $n(E)$ versus E picture which thus emerges is one in which a conduction band (centered about E_F) and a valence band are separated by a minimum in $n(E)$ (see Fig. 2-5(b)). The low-lying band represents covalent bonding which, being highly directional, is responsible for local lattice stiffening in the dilute alloy and bulk stiffening — even brittleness — in the intermetallic compound. Although somewhat more refined, the conclusions arrived at by LYE regarding the electronic structures of intermetallic compounds of Ti with the elements C, N, and O, based on the results of band-structure calculations for TiC, are also in general agreement with this picture [LYE66]. Solution strengthening in ternary solid solutions can be thought of as a two-stage application of this model: strengthening is due not only to the Ti/SM$_1$ and Ti/SM$_2$ interactions — in which case a mixture rule (weighted average) would be obeyed — but also to SM_1/SM_2 interactions which (for a given total solute concentration) allow the strength of a Ti-SM_1-SM_2 alloy to be greater than that of the stronger of Ti-SM_1 and Ti-SM_2. *Fig. 5-10* is an example of this synergistic effect.

(b) Interstitial-Atom Strengthening. CONRAD and co-workers have conducted an extensive investigation of the influences of the so-called "interstitial elements", C, N, and O, on the plastic properties of Ti [CON67, CON70, CON75, CON81][SAR72][OKA73][TYS75]. A detailed study of the influence of N, in particular, was undertaken by OKA et al. [OKA73], whose results were analyzed using the conventional activation-energy approach. An intercomparison of the effects of C, N, and O is presented in [CON75], while the entire subject has been reviewed in [CON81]. The influence of Si has been investigated by FLOWER et al. [FLO73], who also adopted an activation-energy approach when analyzing their temperature-dependent strengthening data.

Although CONRAD et al. have frequently justified their results in terms of conventional lattice-defect theory (including the size-misfit and modulus-defect formalisms), they have gone on to consider the effects of "chemical interaction" between the solute and solvent atoms. In doing so, the interaction mechanism was deduced, with the aid of atomic-orbital theory, to take the form of covalent bonding between the interstitial atom and the matrix. The approach adopted by CONRAD et al. [SAR72][CON75] represents a satisfactory unification of conventional and electronic theories, in that size-misfit and modulus-defect were regarded by them as manifestations of the underlying electronic interaction. The word "chemical" invokes "alloy chemistry" with its thermodynamic (e.g., heats of formation

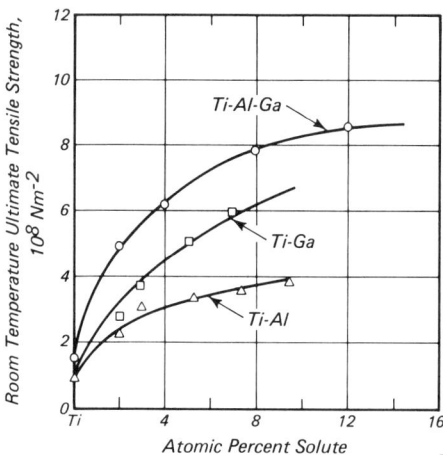

Fig. 5-10. Solid-solution strengthening of Ti with binary and ternary additions of Al and Ga [COL75a].

of the appropriate intermetallic compounds) as well as electronic implications (e.g., electronegativity difference, covalent bonding, etc.). No doubt the calculations by LYE referred to in the previous subsection could provide a useful theoretical basis for this model of local-covalent-bond solid-solution strengthening.

The picture of interstitial-atom hardening that emerged, especially from the paper by SARGENT and CONRAD [SAR 72], was formally indistinguishable from the substitutional strengthening model of COLLINGS and GEGEL [COL 75ª]. If differences do exist between interstitial and substitutional solid-solution strengthening, they are quantitative rather than conceptual and are, therefore, to be found in: (i) the energies of the respective covalent bonds, and (ii) the diffusivities of the dissolved elements.

A comparison of the hardening rates of simple metals and interstitial elements when dissolved in Ti is presented in *Tables 5-3(a)* and *5-3(b)*.

(c) Solution Strengthening by Transition Metals. When transition elements, especially "nearby" ones (referring to the periodic table), are dissolved in Ti, the perturbation of electron states characteristic of the presence of simple metals and interstitials (so-called "s, p elements") does not take place. To a first approximation, the Ti-TM alloy may be regarded as a *new transition metal* with properties — particularly lattice stability — appropriate to the average group number or electron/atom ratio. Although some degree of solution-strengthening is inevitably contributed by the presence of transition metals in β-phase solid solution, the dominant strengthening mechanisms in such alloys are precipitational effects (to be considered below). Of course, in small amounts TM solutes actually *lower* the modulus of Ti, see *Fig. 5-5*. According to *Table 5-4*, the solid-solution strengthening capacity of transition metals is on the average about 30 MN m^{-2} per wt%. That of the "β-isomorphous" early transition elements is about 20 MN m^{-2} per at.%; this is equivalent to a hardening rate of about 6 kg mm^{-2} per at.%* — somewhat smaller than those of the substitutional α stabilizers, *Table 5-3(a)*, and very much smaller than those of the interstitial elements, *Table 5-3(b)*. For this reason, transition elements are regarded as "β stabilizers" and precipitation hardeners rather than solid-solution strengtheners.

(d) Conclusion. Transition elements are stabilizers of the β phase in Ti rather than strengtheners of it. As for the α stabilizers, it may be concluded that there is no conceptual difference between the ways in which the "substitutional" elements and the "interstitial" elements solution strengthen Ti. In both cases the mechanism consists of the establishment

---

*Assuming that $H_V = 3Y$ — see Sect. 5.10.

Table 5-3. Solid-Solution Hardening of Titanium by Simple Metals and by Interstitial Elements [COL84]

Alloying addition	Concentration range, at.%	Condition	Law*	Slope, b, kg mm⁻² at.%⁻¹ᐟ² or kg mm⁻² at.%⁻¹	Intercept, a, kg mm⁻²	Correlation coefficient, %	Hardening rate, $\partial H_V/\partial c$, kg mm⁻² at.%⁻¹ At 0.1 at.%	At 1.0 at.%
(a) Simple metal additions								
Al	0–10	100h/850°C/IBQ	c	15	102	99.6	—	15
Ga	0–5	As-cast	c	24	108	99	—	24
Si	0–2	1h/1000°C/IBQ	c	39	125	83	—	39
Ge	0–5	1h/1000°C/IBQ	c	33	120	98	—	33
Sn	0–7	As-cast	c	24	112	99	—	24
(b) Interstitial-element additions								
B	0–0.2	120h/800°C/IBQ	$c^{1/2}$	218	110	92	344	109
C	0–0.5	120h/800°C/IBQ	$c^{1/2}$	170	104	99.9	269	85
N	0–5	120h/800°C/IBQ	$c^{1/2}$	239	98	99.8	378	120
O	0–3	120h/800°C/IBQ	$c^{1/2}$	194	100	99.9	307	97

*Data fitted to either $H_V = a + bc$ or $H_V = a + bc^{1/2}$.

**Table 5-4. Solid-Solution Strengthening of Titanium
by Transition Elements [HAM78]**

Solid-solution strengthening rate, 10^6 N m^{-2}	Solute element							
	V	Cr	Mn	Fe	Co	Ni	Cu	Mo
Per wt.%	19	21	34	46	48	35	14	27
Per at.%	20	23	39	54	59	43	18	54

of local covalent-like bonds whose strengths may be gauged by any one of several measures of atomic-interaction strength, such as electronegativity difference, heats of solution, heats of formation of intermetallic compounds, and so on. The difference between interstitial- and substitutional-element strengthening is principally one of degree. Taking O and Al as examples, the rates of hardening at the 1 at.% level are in the ratio 6.5:1. The heats of formation of the equiatomic compounds, TiO and TiAl, are in exactly that same ratio [KUB56, pp. 160 and 41, resp.], while the ratio of the GEGEL-calculated electronegativity differences between O and Ti and Al and Ti, respectively, is 6.6:1 [COL75ª]. Further comparisons of this kind are listed in *Table 5-3*.

Solution strengthening by interstitial atoms is chiefly of importance at low temperatures where their diffusivities are low. The same is evidently true, but to a lesser extent, for substitutional (SM) solution strengtheners since, as is well known (cf. *Fig. 5-1* and *Fig. 5-2*), the strengths of the α-solid-solution alloys decrease rapidly with increasing temperature. If strength must be maintained to high temperatures, it is necessary to introduce precipitation strengthening in the form of α_2-phase particles or the numerous microstructural effects available with $\alpha + \beta$ alloying.

5.8.2 Microstructural Strengthening

The temperature range within which solid-solution strengthening is fully effective does not extend far above room temperature. Thus, as indicated in *Fig. 5-1*, the ultimate tensile strength of Ti-5Al-2.5Sn drops 50% upon warming from 20 K to room temperature, and then to 60% of its room-temperature value upon further heating to 370°C. As a consequence, alloy designers have turned to microstructural effects, including precipitation, to extend the temperature range of Ti-base alloys. Since very high-temperature service is equivalent to long-time aging, the most stable precipitation-strengthened, heat-resistant alloy is one that is in thermodynamic equilibrium at the service temperature. For this reason, considerable hope was attached to the possibility that the α_2 precipitate in Ti Al would turn out to be a successful dispersion hardener, comparable in its properties to the γ' phase in Ni-base superalloys. Unfortunately, this did not turn out to be the case, although the possibility of reducing the embrittlement problems associated with α_2-Ti_3Al has been explored [Hoc73][Ham78]. At the other extreme, highly alloyed all-β alloys are not only far removed from Ti-rich alloys in strength/weight ratio and other important "Ti-alloy" characteristics, but they are also single-phase (i.e., precipitate-free) when in thermodynamic equilibrium at high temperatures. Consequently, the tendency has been to turn to $\alpha+\beta$ alloys, and near-α ones at that, for high-temperature service. As indicated in the introduction to this section, when strengthening is due primarily to the presence of a complicated microstructure involving networks of interfaces between finely divided α and β phases, and precipitates within them, the categorization of strengthening sources is generally treated phenomenologically (i.e., in the form of a description of the features present) rather than mechanistically. It is only fair to state, however, that whenever possible serious attempts are made to analyze the strengthening processes in terms of well-established mechanisms such as: (i) direct dislocation-particle interaction, (ii) coherency strains between α and β phases, and (iii) coherency strains between the precipitate and the matrix. The work of Rhodes and Paton [Rho77] is an excellent example of this.

It is clear, in the light of the foregoing remarks, that many of the sections of Chapter 7 ("Aging") could be re-cast in the form of a discussion of "Precipitate Strengthening". There is no need to do this here; however, a summary will be presented, based on the contents of earlier chapters, of the precipitates which have been induced to form in $\alpha+\beta$ and β alloys during attempts to increase their strengths and service-temperature ranges while retaining their ductilities.

Solute-lean β phases when quenched yield a martensitic structure designated α', $\alpha''_{lean}+\alpha''_{rich}$, or α'', depending on the starting composition (see for example Sect. 4.2.5(b)). During aging, all forms result in β-phase precipitation. Isothermal ω forms during the aging of quenched β within an appropriate time-temperature frame, Sect. 7.2. This is responsible for an increase in strength at the expense of ductility. By administering a short higher-temperature heat treatment, the ω phase can be induced to revert to a lean-β (β') precipitate, stable upon re-cooling. This confers on the alloy a strength greater than that of the original quenched β, but without the brittleness associated with the original ω phase, Sect. 7.5. The β' can also be the source of a fine dispersion of α-phase precipitation, which also strengthens the β alloy without embrittling it, Sect. 7.5. In some technical alloys, the $\omega+\beta\rightarrow\beta'+\beta$ reaction may not proceed to completion within a reasonable time; the remaining ω phase then contributes strength while the increased volume-fraction of β phase ensures ductility, Sect. 7.14.1. Various heat treatments, administered to β alloys with and without cold work, are possible. Their results have also been considered in Sect. 7.14.1. The heat treatment of $\omega+\beta$ alloys, at temperatures above the ω-reversion limit, will cause α-phase precipitation to take place in the vicinity of the ω sites, Sect. 7.4.3. This fine dispersion of α phase which results is also an effective strengthener. Heat treatment of metastable β just outside the $\omega+\beta$ field will stimulate the phase-separation reaction and result in a modulated β structure, $\beta'+\beta$, Sect. 7.3. Under heat treatment, the β' can also be the site of fine α-phase precipitation, with beneficial results, Sect. 7.4.2. In technical alloys such as β-C, the $\beta'\rightarrow\alpha$ reaction is complicated and in fact will yield two types of α phase, depending on the heat treatment time/ temperature conditions. By adjusting the aging prescription it is possible to "fine-tune" the α-phase morphology; thus as RHODES and PATON [RHO77] discovered, in β-C the best combination of strength and ductility resulted from the presence of large non-coherent so-called "Type-2" α-phase particles, Sect. 7.14.2.

5.9 Hardness

5.9.1 The Vickers Hardness Test

The measurement of hardness is a simple but useful technique for characterizing mechanical properties and investigating phases in quenched and aged alloys. Conventional techniques currently in use for measuring hardness, as well as the history of that test, are fully discussed by HANKE [HAN54]. In the Vickers method, a weighted square pyramid, usually of diamond, is allowed to rest for a specified length of time on a polished surface of the specimen. Since the area of the impression (mean diagonal, d, mm) is proportional to the load (L, kg), a load-independent hardness number can be formed from the quotient L/d^2. According to the Vickers prescription, $H_V = 1.8544 \, L/d^2$. Thus, for example, a 5-kg load resting on a surface of Vickers hardness $H_V = 150$ kg mm^{-2} will produce a 0.25 mm (diagonal) impression (as in the studies of Ti-V and Ti-Nb referred to in *Fig. 5-11*).

Of course, if it is desired to investigate the individual grains of a fine-grain polycrystalline sample, miniaturized versions of the test are needed. Using for example the Leitz Miniloader, loads in the range of 25-100 g produce measurable impressions less than 10 μm across. Since samples mounted and polished for optical metallography are ideally prepared for hardness measurement, the two investigations are frequently associated with each other in studies of precipitation and aging. Although very useful in tracing the course of an aging reaction, for example, the hardness measurement is obviously not capable of identifying the nature of any precipitating phase. This must always be determined by a separate x-ray or TEM investigation, in the absence of which several alternative possibilities, particularly with regard to submicroscopic continuous precipitation processes (such as for example $\beta \rightarrow \beta' + \beta \rightarrow \alpha + \beta$, $\beta \rightarrow \beta' + \beta''$, and $\beta \rightarrow \omega + \beta$, all of which produce hardening), could easily be confused.

With the aid of formulae such as those due to HILL [HIL67, p. 213] and MARSH [MAR64, MAR64[a]], to be discussed below, Vickers hardness can be related to yield strength. This approach enables tests to be carried out on very small and irregular samples whose yield strengths would otherwise elude measurement.

5.9.2 Hardness of Quenched Ti-TM Alloys

Vickers hardness measurements, aided by x-ray diffraction studies, were employed by BAGARIATSKII et al. [BAG59] in their well-known investigation of the limits of the quenched α'-, α''-, and ω-phase regimes in the systems Ti-V, -Nb, -Ta, -Mo, -W, and -Re. The athermal ω-phase regime was in each system characterized by a pronounced local maximum in the hardness ($\Delta H_V \cong 80$ kg mm^{-2}). In another investigation of alloy phases, this time in the Ti-Nb system, GUZEI et al. [GUZ66] have measured hardness in a series of such alloys quenched from equilibrium at various temperatures. RASSMANN and ILLGEN [RAS72] have also investigated hardness as function of composition in Ti-Nb(10-80 at.%) alloy samples, either as-cast or prepared in cold-rolled form. For both series of samples, the hardness increased rapidly with reduction of the Nb content below 30 at.%, as

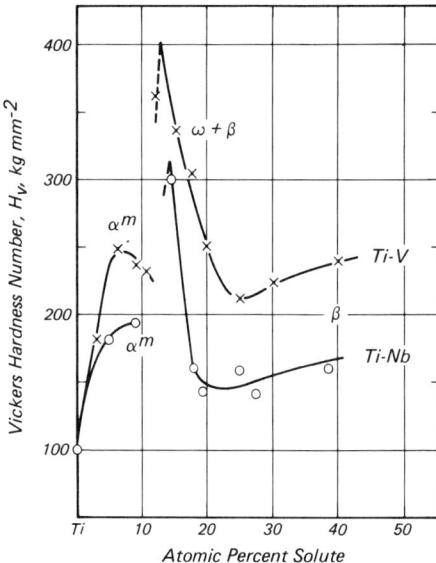

Fig. 5-11. Vickers hardness (5-kg load) of β-quenched Ti-V and Ti-Nb alloys as a function of solute concentration and (by implication) concentration-controlled microstructure [COL84].

expected. The hardness of the cold-deformed alloy was much higher throughout the concentration range, presumably because of work hardening in general and deformation-martensitic transformation (to α'') at the Ti-rich end. The microstructural possibilities in response to 1-h annealing, which leads to a pronounced hardness increase at 300°C, are even more complicated. The frequent attempts which have been made to explore such deformation, precipitation, and aging effects using only hardness as a tool exceeded the capabilities of the technique.

The results of some recent studies of hardness in Ti-V and Ti-Nb alloys are presented in *Fig. 5-11*. The alloys of both systems are seen to become extremely hard as the solute concentration decreases below about 13 at.% ($e/a = 4.13$). In contrast to earlier practice [BAG59][GUZ66], no attempt is made here to generate a smoothly rounded maximum from the low concentration data (see *Fig. 5-5* and *Fig. 5-6*). The discontinuity shown, which is in keeping with the behavior of physical property data such as magnetic susceptibility, electronic specific heat coefficient, and Debye temperature, expresses the strong hardening associated with a high volume-fraction of athermal ω phase; to the left of the discontinuity the alloys have transformed martensitically.

5.9.3 Influence of α Stabilizers on the Hardness of Ti and a Typical β-Isomorphous Ti-TM Alloy

Sect. 5.8 has discussed the principles whereby Ti becomes solution strengthened by: (i) the interstitial elements, (ii) the substitutional α stabilizers, and (iii) other transition metals. An extension of the same principles to the strengthening of β-Ti-TM alloys has been offered in [GEG73a] and [COL75a]. The known relationship between strength and hardness (see Sect. 5.10) has enabled intercomparisons to be made between the rates of strengthening produced by these three classes of solute when dissolved in Ti and β-Ti-TM. Some experimental results are listed in *Tables 5-3, 5-4,* and *5-5*, and depicted in *Fig. 5-12*. Solute species selected for the hardness test were: (i) the interstitial elements B, C, N, and O, (ii) the valence-3 simple metals Al and Ga, and (iii) the valence-4 metals and semimetals Si, Ge, and Sn [COL84]. In presenting for comparison the hardening of a representative β-Ti-TM alloy by some of these same elements, the availability of data dictated the selection of Al and Ge as solutes and Ti-50Nb as the β-Ti-TM solvent, *Table 5-5*. Finally, as indicated in that table, the hardening produced by Zr when dissolved in a β-Ti-TM base was also investigated.

In least-squares fitting the H_V versus concentration (c) data it was noted that hardening by the interstitial elements closely followed a $c^{1/2}$ law (correlation coefficient in most cases >99%), *Table 5-3(b)*. With the elements listed in *Table 5-3(a)*, the hardening rates were sufficiently low

Table 5-5. Solid-Solution Hardening of β-Ti-50Nb by Simple Metals

Alloying addition	Concentration range, at.%	Condition	Law*	Slope, b, kg mm^{-2} at.%$^{-1}$	Intercept, a, kg mm^{-2}	Correlation coefficient, %	Hardening rate, $\partial H_V/\partial c$, at 1 at.% level, kg mm^{-2} at.%$^{-1}$	References
Cu	0-5	3h/1000°C/WQ	c	14	193	99.5	14	[LOH71]
Ge	0-4	3h/1000 or 1080°C/WQ	c	14	165	93	14	[HEL71]
Al	0-2.5	Cold deformed plus 34h/400°C	c	19	195	99	19	[ZWI70]
Ge	0-2	Cold deformed plus 34h/400°C	c	60	198	99	60	[ZWI70]
Zr	0-6	Cold deformed plus 34h/400°C	c	3.7	198	99.9	3.7	[ZWI70]

*Data fitted to $H_V = a + bc$.

that a linear relationship adequately described the data. *Table 5-3* also shows that at the 0.1 at.% level the instantaneous hardening rate of previously unalloyed Ti by interstitials (viz, 269-378 kg mm^{-2} per at.%) is up to twenty times greater than that by, say, Al, or up to nine times faster than that produced by Si. The data of *Table 5-3(a)* and *Table 5-5* together indicate that Al hardens Ti at about the same rate that it does Ti-50Nb (15 and 19 kg mm^{-2} per at.%, respectively). Finally, the above-mentioned tables and *Fig. 5-12* demonstrate the relatively insignificant influence that Zr exerts on the hardness of Ti-50Nb.

Numerical and graphical comparisons, such as those of *Tables 5-3, 5-4,* and *5-5* and *Fig. 5-12*, serve to illustrate that the solid-solution hardening of an α-Ti or β-Ti-TM base is subdivisible into three regimes according to whether the rate of hardening is "low" (neighboring transition elements such as Zr), "moderate" (substitutional simple metals or semimetals such as Al, Sn, Si, and Ge), or "rapid" (the interstitial elements B, C, N, and O). Strengthening mechanisms associated with these three classes of solute have been described in Sect. 5.8.

Again anticipating the results of a discussion to be presented below concerning the proportionality between yield strength and hardness, it is expected that the addition of interstitial elements to Ti (and to Ti-TM alloys) will increase yield strength (at the 0.1 at.% addition level) at the rate of about 30% per 0.1 at.%. On a wt.% basis the strengthening rate due to oxygen is, of course, even greater than this.

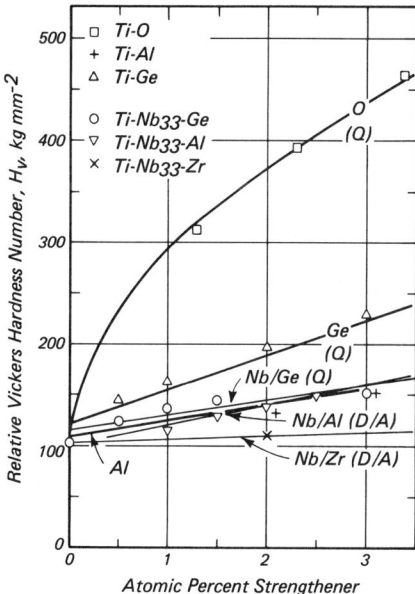

Fig. 5-12. Influence of various interstitial, substitutional-SM, and substitutional-TM additions on the Vickers hardness of Ti and Ti-50Nb in the conditions quenched (*Q*) and deformed-plus-aged (*D/A*). The ternary data have been adjusted to $H_V = 103$ kg mm^{-2} at zero solute concentration. *Data sources*: binary alloys [Col84], Ti-Nb-Ge [Hel71], Ti-Nb-Al/Ge [Zwi70].

5.10 Theoretical Relationships Between Hardness and Strength

A pair of complementary relationships between the hardnesses and strengths of metals have been developed by HILL [HIL67] and MARSH [MAR64, MAR64a]. An excellent review of these theories has been presented by DAVIS [DAV75] in connection with a study of the hardness/yield-strength ratio in metallic glasses.

5.10.1 Hill's Theory

The model assumed by HILL [HIL67] pictured an elastically rigid, yet plastic, body with the material displaced by the indentor squeezed up into a rim around the edges of the imprint. It led to the commonly observed relationship $H_V \cong 3Y$ between the hardness, H_V, and the yield strength, Y, when expressed in the same units (usually kg mm^{-2}).

5.10.2 Marsh's Theory

The theory of MARSH, on the other hand, pictures a cavity being pushed into an elastic body. It results in the semiempirical formula

$$H_V/Y = 0.28 + 0.6\, B \ln Z \ , \tag{5-25}$$

where

$$B \equiv 3/(3 - l) \ ,$$

$$Z \equiv 3/(l + 3m - lm) \ ,$$

with

$$l \equiv (1 - 2\nu)(Y/E) \ ,$$

$$m \equiv (1 + \nu)(Y/E) \ ,$$

ν being Poisson's ratio and E the Young's modulus.

5.10.3 Relationship Between the Models

The HILL and MARSH models represent extreme limits of the relative magnitudes of E and Y. It turns out that if $133(Y/E)$ is "small" (<1) we have the plastic/elastically rigid situation discussed by HILL, whereas if $133(Y/E)$ is "large" (>1) the situation is identifiable as the unyielding (non-plastic)/elastic condition required by MARSH. The crossover value of Y/E is $1/133$.

With the aid of the following approximations, the MARSH equation becomes much simpler to apply in practice: For many polycrystalline metals, ν may be replaced by 0.3 and Y/E is generally found to be of the order 10^{-2}.* Using this information:

$$l = (1 - 2\nu)(Y/E) \cong 0.4 \times 10^{-2} \ , \qquad (5\text{-}26)$$

hence

$$B \cong 1$$

Similarly,

$$m = (1 + \nu)(Y/E) \cong 1.3 \times 10^{-2} \ .$$

Now by definition

$$Z = 3/(l + 3m - lm) \ ;$$

but if

$$l \cong 0.4 \times 10^{-2} \ ,$$

and

$$3m \cong 4 \times 10^{-2} \ ,$$
$$lm \sim 10^{-4} \ ,$$

and is negligible compared to $l + 3m$. Thus,

$$Z \cong 3/(l + 3m) \ ,$$

which from the definitions of l and m becomes

$$Z \cong E/1.43Y \ . \qquad (5\text{-}27)$$

Inserting (5-26) and (5-27) into (5-25) yields finally

$$H_V/Y \cong 0.065 + 0.6 \ln(E/Y) \ , \qquad (5\text{-}28)$$

---

*In Ti-Nb(40 at.%), for example, $Y = 8.8 \times 10^8$ N m^{-2} [Koc77] and $E = 8.0 \times 10^{10}$ N m^{-2} [Led81].

applicable, as mentioned above, for $E/Y < 133$ which according to numerical solution of the equation corresponds to $E/H_V < 44.3$.

5.11 Application of the Marsh Formula to the Determination of the Yield Strength of a Wire

The microhardness, H_V, is determined upon a metallographically mounted cross section of the wire. The Young's modulus, E, is determined using either a sonic or an ultrasonic sound-velocity measurement. If $E/H_V > 44.3$, the HILL relationship $Y = H_V/3$ is immediately accepted. Otherwise, the MARSH approach is adopted and Eqn. (5-28) solved transcendentally for E/Y, hence Y. The Newton-Raphson method, which when applied to Eqn. (5-28) converges very rapidly, is well suited for this purpose.

5.11.1 Newton-Raphson Solution of the Simplified Marsh Equation

(a) **Direct Method.** If X_i approximately satisfies $F(X_i) = 0$, then a better root is X_{i+1}, where

$$X_{i+1} = X_i - \frac{F(X_i)}{F'(X_i)} \ . \tag{5-29}$$

By iterating Eqn. (5-29) a few times, excellent solutions can be obtained.

In the present case, values for H_V and E having been obtained, the root Y of the function $F(Y) \equiv 0.065 - H_V/Y + 0.6\,ln(E/Y)$ is required. This is quickly obtained by substituting $y = Y^{-1}$ and iterating

$$y_{i+1} = y_i - \frac{0.065 H_V y_i + 0.6\,ln(E y_i)}{-H_V + 0.6/y_i}$$

(b) **Alternative (Trial-and-Error) Method.** A trial-and-error method commences with

$$H_V/Y \cong 0.065 + 0.6\,ln\,(E/Y) \ . \tag{5-28}$$

Let

$$E/H_V \equiv R \ , \tag{5-30}$$

then

$$H_V/Y = 0.065 + 0.6 \, ln \, R + 0.6 \, ln \, (H_V/Y) \ ,$$

so that the equation to be solved by trial-and-error for H_V/Y is:

$$H_V/Y - 0.6 \, ln \, (H_V/Y) = R' \ , \tag{5-31a}$$

where

$$R' \equiv 0.065 + 0.6 \, ln \, R \ . \tag{5-31b}$$

5.11.2 Graphical Solution of the Marsh Equation

Fig. 5-13 has been constructed from Eqns. (5-28), (5-30), and (5-31). Below the HILL lower bound ($E/H_V = 44.3$), values of H_V/Y or E/Y corresponding to measured values of E/H_V can be simply read from the graph.

5.12 Interrelationships Among Hardness, Young's Modulus, and Yield Strength in a β-Isomorphous Alloy Sequence: A Case Study

Ti-Nb, a typical β-isomorphous system, has been selected for this study. READ [REA78] has measured the hardness and yield strengths of the alloys Ti-Nb(30, 39, and 50 at.%) in the "annealed" and "processed" conditions as defined in *Table 5-6*, in which all the mechanical-property data are summarized. The moduli of the "annealed" alloys were also reported, but not those of the "processed". To replace the missing data, values of E were deduced by extrapolation and interpolation of the dynamic-modulus results of ALBERT and PFEIFFER [ALB72] for Ti-Nb(34 at.%) and Ti-Nb(44 at.%).

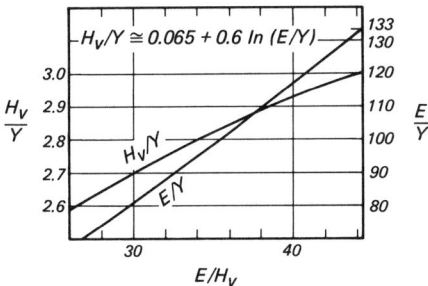

Fig. 5-13. Parametric representation of the relationship $H_V/Y \cong 0.065 + 0.6 \, ln(E/Y)$, Eqn. (5-28).

Table 5-6. Estimation of Yield Strength from Measured Hardness Data Using the Models of HILL and MARSH Where Applicable: A Case Study of Three β-Ti-Nb Alloys

Alloy	Experimental input* Hardness, H_V, kg m^{-2}	Modulus, E, 10^{10} N m^{-2}	Modulus, E, 10^6 g mm^{-2}	E/H_V	Model regime based on the magnitude of the E/H_V ratio	Calculation of yield strength Hardness/ strength ratio from application of model, H_V/Y	Yield strength calculated from hardness, Y_{calc} kg mm^{-2}	Y_{calc}, 10^8 N m^{-2}	Measured yield strength, Y_{meas}, 10^8 N m^{-2}	Deviation, $100\,\Delta Y/Y_{meas}$, %
Condition: "Annealed"										
800°C and water quenched										
Ti-Nb(30 at.%)	145	6.6	6.7_3	46	HILL	3.0	48	4.7	4.5	4
Ti-Nb(39 at.%)	150	7.8	7.9_5	53	HILL	3.0	50	4.9	4.6	7
Ti-Nb(50 at.%)	147	8.5	8.6_7	59	HILL	3.0	49	4.8	4.1	17
Condition: "Processed"										
Ti-Nb(30, 39 at.%): (800°C/WQ)/extruded 86%/(48h/375°C)/drawn 56%										
Ti-Nb(50 at.%): As-received cold-worked condition										
Ti-Nb(30 at.%)	293	7.1_5 *	7.2_9	25	MARSH	2.5_6	114	11.2	10.3	9
Ti-Nb(39 at.%)	174	7.8_8	8.0_4	46	HILL	3.0	58	5.7	7.0	19
Ti-Nb(50 at.%)	152	8.7_7	8.9_4	59	HILL	3.0	51	5.0	5.2	4

*All input data and measured yield strengths are due to READ [REA78] except the enclosed modulus data, which have been derived from the dynamic-modulus results of ALBERT and PFEIFFER [ALB72] for Ti-Nb(34, 44 at.%) in the condition 95 % cold worked plus 40h/400°C.

The first step in deriving Y from the measured H_V and E values is to determine whether E/H_V (as listed in the 5th column of *Table 5-6*) is greater than or less than 44.3. In all cases but one, $E/H_V > 44.3$, enabling Y to be obtained from the HILL relationship $Y = H_V/3$. The last three columns of *Table 5-6* compare the calculated and measured values of Y. In most cases the agreement is remarkably good, the sole exception being Ti-Nb(39 at.%)("processed"), whose calculated yield strength is 19% too low, a deficiency possibly traceable to a low measured hardness number.

5.13 Tensile Strengths of Some Commercial Titanium Alloys

In selecting a Ti alloy for a given application, strength (and indeed the other mechanical properties) is not of course the only, or even the most important, consideration. Only strength properties are presented and discussed in this section. Other important properties which govern the selection of a Ti alloy for a given application are *weldability* and *heat resistance* (the latter having to do with the rate at which the strength decreases with temperature in the elevated-temperature regime, e.g., 600-1000°F). Of course, the properties of even a single alloy type can be altered by heat treatment (in the case of the $\alpha + \beta$ and metastable β classes) and by slight variation or "fine-tuning" of its composition (e.g., at least half-a-dozen varieties of "Ti-6-4" can be produced in this way).

Ti-6Al-4V is the most commonly used Ti-base material (including unalloyed commercial Ti). As indicated above, it is produced in a number of varieties by slight alteration of the basic composition; it is also produced in numerous heat-treatment conditions in order to optimize its performance for particular applications within a wide range of service conditions. Ti-5Al-2.5Sn is a non-heat-treatable intermediate-strength alloy. The extra-low-interstitial (ELI) grade is especially noted for its excellent properties at cryogenic temperatures; this alloy, which has also been found to possess useful strengths at intermediate-to-high temperatures (up to 700-800°F), is suitable for aircraft engine and frame applications. Ti-3Al-2.5V has a lower strength than both Ti-5Al-2.5Sn and Ti-6Al-4V; its primary use is for aircraft hydraulic tubing. Ti-8Al-1Mo-1V, a near-α $\alpha + \beta$ alloy, has the highest Al content of any of the currently available commercial alloys. A moderately high-strength material, it is used for gas-turbine-engine cases, discs, and compressor blades, as well as other ambient-temperature aircraft applications. Ti-5Al-2Sn-2Zr-4Mo-4Cr (Ti-17) was developed specifically to meet the requirements of gas-turbine-engine discs. Ti-6242 was developed in the mid-1960s as a gas-turbine material capable of service at 1000°F. Ti-6246 was developed in the late-1960s as an improved performance material for use in gas-turbine-engine discs and compressor components.

The room-temperature tensile and compressive strengths of a number of commercial Ti-base alloys are presented in *Table 5-7(a)*. The low-temperature tensile properties of a selected few are presented in *Table 5-7(b)*. For useful extended summaries of the properties and applications of the alloys listed in the tables, a recently published Appendix to the *Structural Alloys Handbook* [STR82] is recommended.

Part 3: Anomalous Plastic Properties

5.14 Normal and Anomalous Tensile Properties

When a metal is stressed uniaxially, say in tension, it extends linearly and elastically for a time, then after passing a yield point begins to elongate plastically, still smoothly but at a much faster rate than before. For most purposes the information describing the results of such a tensile test are: the Young's modulus (E), the 0.2% offset yield strength (Y or $\sigma_{0.2}$), the ultimate strength (Y_{ULT}), the fracture strength (σ_B), and whether or not the fracture was brittle or ductile. These might be referred to as the "normal" tensile properties. All materials possess them; however, close attention to the details of the tensile and/or compression test and its results often reveals what might be termed "anomalous tensile properties". Properties to be defined and discussed below with particular reference to Ti alloys are: serrated yielding, acoustic emission, thermoelasticity, pseudoelasticity, shape-memory effect, and Bauschinger effect. They are illustrated in *Fig. 5-14*. Depending on their purity (interstitial content), phase stability (proximity to a phase boundary), grain size, temperature, strain-rate, and so on, some materials/samples exhibit anomalous properties more strongly than others.

5.14.1 Serrated Yielding

The jagged oscillation in stress at constant strain-rate that characterizes the $\sigma(\epsilon)$ curves of many metals is referred to as serrated yielding. At low temperatures, if the material being tested is susceptible to stress-induced martensitic transformation or twinning, the output trace of a screw-driven or other such extension-controlled tensile machine will exhibit a sharp load drop each time a spontaneous elongation of the sample takes place. At high temperatures, serrated yielding is the manifestation of a stress-strain relaxation oscillation. Microscopically, this consists of the cyclical repetition of a sequence of events in which short bursts of plastic flow are abruptly terminated by rapid-diffusion-induced local solution-strengthening, a phenomenon to which the term "dynamic strain aging" has been

Table 5-7(a). Tensile Strengths of Several Commercial Titanium-Base Alloys: Typical Room-Temperature Values [STR82]

Alloy name	Nominal composition	Condition	Ultimate strength ksi	Ultimate strength 10^8 N m^{-2}	Yield strength ksi	Yield strength 10^8 N m^{-2}	Elongation, %
5-2.5	Ti-5Al-2.5Sn	Annealed ($\frac{1}{4}$-4h)/(1300-1600°F)	120-130	8.3-9.0	115-120	7.9-8.3	13-18
3-2.5	Ti-3Al-2.5V	Annealed (1-3h)/(1200-1400°F)	95	6.5	90	6.2	22
6-2-1-1	Ti-6Al-2Nb-1Ta-1Mo	Annealed ($\frac{1}{4}$-2h)/(1300-1700°F)	125	8.6	110	7.6	14
8-1-1	Ti-8Al-1Mo-1V	Annealed 8h/1450°F	145	10.0	135	9.3	12
Corona 5	Ti-4.5Al-5Mo-1.5Cr	α-β annealed after β processing	140-160	9.7-11.0	135-150	9.3-10.3	12-15
Ti-17	Ti-5Al-2Sn-2Zr-4Mo-4Cr	α-β or β processed plus aged	165	11.4	155	10.7	8
6-4	Ti-6Al-4V	Annealed 2h/(1300-1600°F)	140	9.6	130	9.0	17
		Aged	170	11.7	160	11.0	12
6-6-2	Ti-6Al-6V-2Sn	Annealed 3h/(1300-1500°F)	155	10.7	145	10.0	14
		Aged	185	12.8	175	12.1	10
6-2-4-2	Ti-6Al-2Sn-4Zr-2Mo	Annealed 4h/(1300-1550°F)	145	10.0	135	9.3	15
6-2-4-6	Ti-6Al-2Sn-4Zr-6Mo	Annealed 2h/(1500-1600°F)	150	10.3	140	9.7	11
		Aged	175	12.1	165	11.4	8
6-22-22	Ti-6Al-2Sn-2Zr-2Mo-2Cr-0.25Si	α-β processed plus aged	162	11.2	147	10.1	14
10-2-3	Ti-10V-2Fe-3Al	Annealed 1h/1400°F	140	9.7	130	9.0	9
		Aged	180-195	12.4-13.4	165-180	11.4-12.4	7
15-3-3-3	Ti-15V-3Cr-3Sn-3Al	Annealed $\frac{1}{4}$h/1450°F	115	7.9	112	7.7	20-25
		Aged	165	11.4	155	10.7	8
13-11-3	Ti-13V-11Cr-3Al	Annealed $\frac{1}{2}$h/(1400-1500°F)	135-140	9.3-9.7	125	8.6	18
		Aged	175	12.1	165	11.4	7
38-6-44	Ti-3Al-8V-6Cr-4Mo-4Zr	Annealed $\frac{1}{2}$h/(1500-1700°F)	120-130	8.3-9.0	113-120	7.8-8.3	10-15
		Aged	180	12.4	170	11.7	7
β-III	Ti-4.5Sn-6Zr-11.5Mo	Annealed $\frac{1}{2}$h/(1300-1600°F)	100-110	6.9-7.6	95	6.5	23
		Aged	180	12.4	170	11.7	7

Table 5-7(b). Tensile Strengths of Several Commercial Titanium-Base Alloys: Low-Temperature Values [SAL79]

Alloy	Condition	Test temperature, K	Ultimate strength, 10^8 N m^{-2}	Yield strength, 10^8 N m^{-2}	Elongation, %
Ti-5Al-2.5Sn (5-2.5)	Annealed, normal interstitial	295	8.8	8.6	16
		200	11.0	10.6	14
		77	14.0	13.7	12
		20	16.9	16.8	5.1
	Annealed, extra-low interstitial (ELI)	295	7.6	7.1	17
		200	9.2	8.6	16
		77	12.6	11.9	17
		20	15.4	14.4	15
Ti-8Al-1Mo-1V (8-1-1)	Annealed	295	10.3	9.7	16
		200	11.9	11.3	14
		77	15.6	14.4	13
		20	17.5	16.2	2.4
	Duplex annealed	295	10.2	9.5	15
		200	11.2	10.3	15
		77	14.9	13.4	22
		20	16.9	16.1	1.2
Ti-6Al-4V (6-4)	Annealed, normal interstitial	295	9.9	8.9	12
		200	11.6	10.7	11
		77	15.3	14.3	11
		20	17.9	17.3	2.4
	Solution treated and aged, normal interstitial	295	12.2	11.3	8
		200	13.2	12.8	6
		77	17.6	17.0	5
		20	20.4	19.9	0.7
	Annealed, ELI	295	9.9	9.3	12
		200	11.5	10.9	12
		77	15.1	14.6	10
		20	18.2	17.9	2.9
	Solution treated and aged, ELI	295	11.2	10.6	9
		200	13.2	13.2	7
		77	17.2	16.7	5
		20	19.6	19.6	1.0
Ti-13V-11Cr-3Al (13-11-3)	Annealed or solution treated	295	9.7	9.4	19
		200	12.5	12.2	12
		77	19.5	18.9	2.1
		20	22.6	—	0.5
	Solution treated and aged	295	13.6	12.4	7
		200	15.7	14.7	2.1
		77	16.5	—	0.2
		20	—	—	—

aptly applied. Depending on their composition, the temperature, and their condition, Ti alloys exhibit both classes of serrated yielding.

(a) Serrated Yielding at Low Temperatures. At low temperatures where diffusion rates are low, serrated yielding, when it occurs, is generally a property of the crystal lattice. The low-temperature phenomenon has been well discussed by BASINSKI [BAS57], who, by using Al and Al alloys as samples, disassociated the general effect from both twinning and martensitic transformation. Necessary conditions for serrated yielding in BASINSKI's experiment were (i) a strong negative temperature dependence of the yield stress and (ii) the usual small low-temperature specific heat. Since under such conditions local slip-nucleation is sufficient to trigger the thermomechanical instability, it was regarded as a quite general low-temperature mechanical property [BAS57]. The local mean temperature increase associated with a serration was calculated by BASINSKI to be ~60 K. He also obtained direct experimental evidence for the temperature increase in an Al test piece by noting the transition to the normal state of a Nb wire imbedded axially in it.

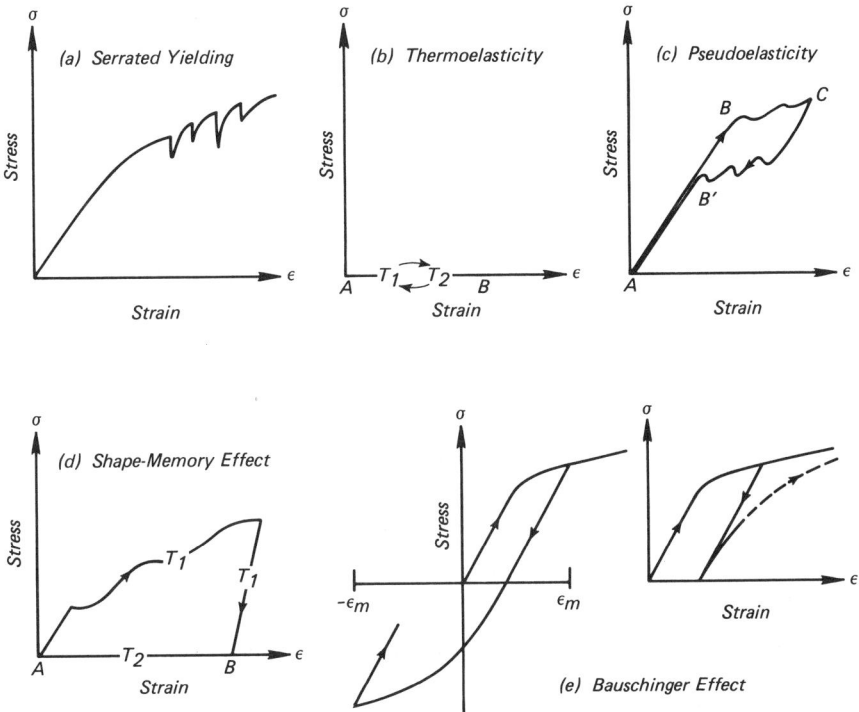

Fig. 5-14. Stress-strain signatures of five anomalous plastic properties.

MOSKALENKO *et al.* [Mos80] have suggested a dislocation mechanism to explain the observed low-temperature ($\lesssim 4.2$ K) serrated yielding in unalloyed Ti (iodide and commercial) and Ti-2.4Zr-1.2Mo. The effect was not attributed to any kind of lattice instability; rather, a relaxation-oscillation mechanism was suggested based on the cycle: *dislocation pile-up/stress increase/dislocation avalanche/local adiabatic heating/load drop* — clearly a low-temperature version of that referred to above.

Serrations have been noted whenever β-Ti-Nb alloys (about 50 wt.%, 34 at.%) were strained at liquid-He temperatures beyond the linear elastic region [Eva73][Hil73ᵃ][Eas75, Eas77][Alb76][Koc77][Pas78]. In a series of tensile tests of Ti-50Nb conducted at various temperatures from 300 down to 4 K, the onset temperature for serrated yielding was found to be about 28 K [Alb76]. EVANS [Eva73], in a search for a mechanism underlying the "training" phenomenon in superconducting magnets, seems to have been the first to apply to Ti-Nb the general concepts of unstable low-temperature plastic flow previously observed in Al alloys. Those experiments, and those of MOSKALENKO *et al.* [Mos80] on unalloyed α-Ti and α-Ti alloys reported above, have demonstrated that no exotic deformation mechanisms were needed, provided that a sufficiently steep negative yield-stress temperature dependence was present. Thus, for an alloy such as Ti-50Nb with inherent lattice instability, manifesting itself as thermoelasticity, pseudoelasticity, or as a shape-memory effect, low-temperature serrated yielding is ensured.

In studies of the deformation and fracture of Ti-50Nb at 4.2 K, HILLMANN attributed the serrations to spontaneous slip taking place on the {110} principal bcc glide planes [Hil73ᵃ]. ALBERT and PFEIFFER [Alb76], in tests of cold-worked (>90%) Ti-50Nb wires possessing the usual ⟨110⟩ texture parallel to the wire axis, noted that the fracture surface was at 45° to the wire axis, and concluded that slip in these samples took place on the {100} planes. A conventional slip-plasticity model was adequate to explain the effects of prior deformation or load alternation. Within such a model, the addition of work-induced defects (work hardening) had the effect of inhibiting slip, thereby moving the serrations up to higher and higher levels of strain, in agreement with observations. Returning to the temperature dependence of the effect, it was mentioned above that with decreasing temperature the instabilities in Ti-Nb appeared for the first time at 28 K. Although this value is in good accord with the extrapolated M_s transformation curve for Ti-Nb (see *Fig. 4-9*), it has been noted [Hil76] that the transition from the "normal" parabolic $\sigma(\epsilon)$ region to the "anomalous" region characterized by intermittent load drops is gradual and not associated with a definite transition temperature. Detailed studies of the temperature dependences of the tensile properties by HILLMANN *et al.* [Hil76], one of the few groups to include 20 K in the measurement program, have shown that the entry into the serrated-yield region is accompanied by several important departures from the behavior expected simply by extrapolating from higher temperature results. The σ_B and $\sigma_{0.2}$ versus T curves are characterized by: (i) considerable scatter in the values of $\sigma_B(T)$, $\sigma_{0.2}(T)$, and $\sigma_{0.01}(T)$ obtained from repeated experiments; (ii) a $\sigma_B(T)$ which passes through a broad maximum in the vicinity of 20 K; (iii) yield strengths, $\sigma_{0.2}(T)$ and $\sigma_{0.01}(T)$, which increase rapidly below about 30 K, but which for cold deformation $\lesssim 90\%$ also maximize at about 20 K; and (iv) elongations, δ_B, which with decreasing temperature drop rapidly from $5 \sim 6\%$ to below 1%. All these effects, which are clearly related to each other and to the serrated-yielding effect in Ti-Nb, warrant further study.

READ [REA 78] has detected serrated $\sigma(\epsilon)$ behavior in Ti-45Nb at 4 K. The appearance of the fracture surface indicated that at least some of the deformation took place by irreversible twinning. EASTON and KOCH have noted serrated yielding in Ti-Nb(36 at.%) wire (0.0105 in.$^{\phi}$) at 4.2 K [EAS75][KOC77] and in a sample of filament (~ 280 μm$^{\phi}$) of comparable composition etched from a commercial multifilamentary composite conductor. In agreement with SCHMIDT [SCH76], they recorded that serrations appeared not only with increasing but also with decreasing load, a reversibility which suggested the operation of a pseudoelastic effect such as the reversible stress-induced martensitic transformation or strain-reversible twinning.

In any investigation of serrated yielding, the possibility that the results are being perturbed by the presence of sample artifacts and the characteristics of the testing machine must be considered. SCHMIDT [SCH77] has considered in some detail the influence of the dynamics of the sample and load train on elongation and load drop during serrated yielding. With regard to the former, he pointed out that the rate of instantaneous *elongation* is determined in part by the inertia of the loading system and is, therefore, apparatus dependent. As for the *load drop*, SCHMIDT noted that during the instant of localized yield the rest of the sample relaxes elastically within a few microseconds, a time negligible compared to the response time of a relatively massive load train which, in effect, remains "rigid" during the process (see Ref. (35) of [KOC77]). Accordingly, it was concluded that under such conditions the magnitude of the load drop should be independent of the apparatus.

(b) Serrated Yielding at High Temperatures. Although the occurrence of dynamic strain aging via solute-atom diffusion is sufficient to ensure serrated yielding, it is not necessarily the only mechanism. Even remote from the low-temperature regime referred to in the previous sub-section, serrated yielding in some alloys has been attributed to lattice instability. For example, β-phase instability was regarded as being responsible for the serrated yielding observed in Ti-3Al-16Mo-10Cr during tensile testing at 200-400°C [Bog73], since the same alloy after $\alpha + \beta$ annealing exhibited no such effect. Similarly β-III [Oht73][Rac75] and some other β-Ti alloys (Ti-15Mo-5Zr, Ti-12Mo-6Sn, and Ti-15Mo-0.2Pd [Oht73]) during testing at $300 \sim 400$°C exhibited serrated yielding, due it was supposed to an isothermal $\beta \rightarrow \omega$ transformation.

Dynamic strain aging is the most frequently cited mechanism of elevated-temperature serrated yielding. It is generally interpreted as the manifestation of a hardness oscillation brought about by the periodic movement of solute atoms sufficiently fast to slow down the moving dislocations with which they become associated [Bri70]. Since interstitials, which occur abundantly in Ti alloys, are the most mobile of the solute atoms, serrated yielding in Ti alloys is usually attributed to them. For example, Garde *et al.* [Gar72] showed that dynamic strain aging in iodide Ti (C, N, O = 78, 6, 63 ppm, resp.) was very much less pronounced than in commercial Ti (C, N, O = 200, 100, 1360 ppm, resp.). In studies of dilute alloys of Ti with Si [Win73], Zr [Sas75, Sas82], and the solutes Hf, Ag, V, In, Sn, Al, Nb, and Ta [Sas82], serrated yielding was again attributed primarily to the presence of interstitial atoms, especially since the amplitudes of the serrations were higher in those alloys which exhibited the highest internal-friction relaxation peaks [Sas82]. The presence of substitutional impurity was not, however, ignored; it was thought to assist in the dynamic strain aging process by distorting the interstitial-atom site [Win73][Sas75]. Furthermore, substitutional atoms can in their own right play a role in dynamic strain aging. Although their diffusivities would normally be too small to satisfy the conditions necessary to establish "atmospheres" about moving dislocations, this is no longer true in the presence of a high density of vacancies such as will become established during plastic flow [Bri70]. Thus, as pointed out by Döner and Conrad [Don75] in a very important paper on elevated-temperature (0.3-0.6 T_m) deformation in Ti-5Al-2.5Sn (0.5 at.% O_{eq}), the dynamic strain aging which was observed in the temperature range $320 \sim 560$°C could very well have been due to the diffusion of the Al and/or the Sn atoms under the conditions of the experiment.

5.14.2 Acoustic Emission

(a) Introduction. In a comprehensive review of proposed and confirmed sources of emission, Jaffrey [Jaf79] has found it convenient to deal with acoustic emission in terms of: (i) phase transformation, (ii) dislocation motion, and (iii) fracture mechanics. The first of the above topical areas included elastic twinning, deformation twinning (e.g., the "cry" of Sn), and martensitic transformation. BASINSKI, for example, during the low-temperature studies referred to earlier, discovered that the martensitic transformation of austenitic stainless steel at temperatures below 45 K was accompanied by tensile load drops and loud clicks [BAS57]. With regard to dislocation motion as a possible source, it has been pointed out that whereas acoustic energy is expected to be emitted from individual moving screw and edge dislocations, at 10^{-23} J per event it was below the detectibility limit of contemporary equipment ($\sim 10^{-16}$ J, 1979). Similarly, although sound was also expected to be emitted during the annihilation of a single dislocation at a free surface, or of a pair of colliding oppositely signed dislocations, the per-event energy was again calculated to be below the detectibility threshold [JAF79]. On the other hand, should several hundreds or thousands of dislocations move cooperatively, as in an avalanche, their combined acoustic output would rise above the instrumentational threshold. Thus, acoustic emission can occur in association with microplastic flow (see Ref. (9) of [SCH77^a]). It has also been detected in association with the normal plastic deformation of Cu at strain levels above 0.3%, during which noise in the 100-300 kHz frequency range was emitted continuously during straining [PAS78]. In the area of fracture mechanics, acoustic emission has been postulated and/or detected in connection with: (i) microcracks forming ahead of subcritical crack tips, (ii) dislocation breakaway at the tips of cracks, and (iii) void nucleation and growth.

(b) Acoustic Emission in Ti Alloys. Acoustic emissions from Ti alloys, as they have been studied and reported, fall within the general categories (i) and (ii) referred to above.

The mechanical stability of Ti-50Nb, because of its use in the multifilamentary composite conductors of large superconducting magnets, has received considerable attention. Serrated yielding in that alloy has already been referred to. Audible clicking associated with load drops were occasionally noted by REED et al. [REE77] during the tensile testing of a Ti-Nb superconductor at 4.0 K. EASTON and KOCH [EAS75], in a more extensive study of audio-frequency emission during the alternate loading and unloading of a 2-mm$^\phi$ tensile-test sample of Ti-55Nb noted that: (i) while emission occurred almost immediately upon application of the stress at 4.2 K, none was detected at 77 or 300 K; (ii) emission took place upon both loading and unloading and *bursts of noise (clicks) preceded any load drop that occurred*; (iii) emission took place *almost immediately* upon application of stress at 4.2 K; and (iv) during repeated loading, the emission *increased in amplitude and frequency* when the immediately preceding stress level was exceeded. It is clear from a close examination of these results and an intercomparison of the italicized statements that pseudo-elastic serrated yielding is a sufficient but not necessary condition for acoustic emission in Ti-Nb at low temperatures.

These results were amplified and clarified in the reports of a subsequent series of experiments by PASZTOR and SCHMIDT [SCH77ª][PAS78]. In studies of ultrasonic emission (100-300 kHz) during the straining of Ti-50Nb wire at 4.2 K, the above authors were able to resolve the effect into reversible and irreversible components depending on whether or not emission would take place during the unloading half-cycle. The threshold for the *reversible* component was a strain level of about 0.4%, which also marked the onset of hysteretic stress-strain behavior of the type referred to as pseudoelastic by EASTON and KOCH [EAS75]. The *irreversible* component was present throughout the entire stress range, and was detectable even at room temperature using 6 db of additional amplifier gain. Noise output took place in bursts of some 100 μs time duration, commencing at extremely low levels of strain (~0.005%). In repeated loading to successively higher levels of strain, the irreversible noise signals (and the only noise generated at all for strains ≳0.3-0.4%) began only when the previous stress level was exceeded. The research of KOCH and EASTON [KOC77] [EAS75] and its extension by PASZTOR and SCHMIDT [SCH77ª][PAS78] has demonstrated that in Ti-50Nb at 4.2 K both irreversible serrated yielding (accompanied by its acoustic emission), which began at about 1% strain [SCH77ª], *and* pseudoelasticity (accompanied by stress-strain hysteresis and *reversible* acoustic emission [EAS75]), beginning at about 0.4% strain [SCH77ª][EAS75], could take place at stress levels as low as ~50 ksi. Furthermore, the occurrence of additional *irreversible* acoustic emission, manifesting itself as steady streams of bursts beginning at extremely small levels of strain (≳0.01%), was thought to be symptomatic of movement at the atomic level.

VODOLAZKY *et al.* [VOD80] have used acoustic emission to study deformation processes in Ti alloys. The material selected for testing in tension as well as in compression at 850°C was Ti-6Al-2Zr-1.5V-1.5Mo. As indicated above, plastic deformation of metals is known to be accompanied by the continuous generation of microcracks, which appear even during the initial stages of plastic deformation. By monitoring the acoustic emission associated with microcrack formation, an early warning of impending fracture, even during high-temperature testing, could be obtained.

5.14.3 Thermoelasticity

The thermoelastic martensitic transformation, *Fig. 5-14(b)*, is one in which a continuous transformation takes place as the temperature is lowered [DEL 74]. In contrast to the more usual mode of the martensitic transformation, the thermoelastic kind is not accompanied by the sudden appearance, or bursts, of groups of platelets. The effect *is* reversible, the volume fraction of martensite decreasing continuously to zero as the temperature is restored to its original value. Implicit in the extrapolated martensitic transformation diagram of *Fig. 4-9* is the suggestion that Ti-Nb alloys of compositions close to those used in commercial superconductors should be thermoelastic in the He-temperature regime. Although an intriguing possibility, there is at present no direct microscopic evidence for such a low-temperature reversible transformation. Indeed, indirect evidence refutes such conjecture, and favors instead the reversible athermal ω-phase transformation. Comparative studies of the room-temperature and He-temperature physical properties of Ti-TM alloys indicate that the composition of the martensitic-phase boundary remains fixed as the temperature decreases below room temperature. Thus, whereas a reasonable extrapolation of the M_s curve for Ti-V [ZWI75, p. 174] passes through, for example, the point ($-200°C$, 19 at.% V), a comparison of the room-temperature magnetic susceptibility composition dependence with that of the superconducting transition temperature shows that the compositional range of the martensitic regime is no greater at He temperatures than it is at 298 K—certainly less than 12 at.% V in both cases [COL75[b], COL75[c], COL82[b]] (see also [COL80]). On the other hand, there is abundant irrefutable direct and indirect evidence for the reversibility of athermal ω phase in alloys of this type. Having rejected, with respect to Ti-TM alloys, reversible low-temperature martensitic transformation in favor of ω phase, one can only speculate at this stage as to *its* ability to lead to a measurable thermoelastic effect. A simple geometrical calculation leads to the prediction that a net 3% contraction takes place perpendicular to the $(\bar{1}12)_\beta$ plane during the $\beta \to \omega$ transformation [LUH70]. Since this is able to be accommodated by coherency strains [HIC69, HIC69[a]], a thermoelastic effect via this mechanism is also unlikely.

5.14.4 Pseudoelasticity

The pseudoelastic effect is illustrated schematically in *Fig. 5-14(c)*. In it the sample, after a period of elastic extension to B, is shown deforming in an apparently plastic manner to C, and upon unloading returning spontaneously to B' with net dissipation of energy, then elastically back (or almost so) to the initial state, A [DEL74]. In this mechanical analog of the thermoelastic effect, martensitic transformation proceeds continuously with increasing applied stress and is reverted continuously when that stress is removed. At sufficiently low stress levels, the sample exhibits normal elastic properties. The apparent plastic deformation is thus the result of a (reversible) stress-induced martensitic transformation. The behavior in the plastic regime suggests that the effect would be more appropriately entitled "pseudoplasticity"; the term "pseudoelasticity" arises from the return of the sample to its original state after the removal of the stress, as in normal elastic behavior.

There is of course abundant evidence for stress-induced martensitic transformation in Ti-TM alloys. Sect. 4.2.4 has shown how the transformation can be triggered by the application of additional strain in the vicinity of a strain spinodal defined by $\partial^2 g / \partial \epsilon^2 = 0$. That is to say, in a lattice which is prone to transform—a situation signalled for example by a low value of the elastic shear modulus $C' = (C_{11} - C_{12})/2$—the application of a few percent strain is sufficient to initiate the martensitic reaction. The special characteristic of pseudoelasticity is, of course, the reversible nature of the transformation. As explained by DELAEY et al. [DEL74], pseudoelasticity can also occur in the absence of martensitic phase change by some form of mutual accommodation between pairs of twinned structures.

(a) Pseudoelasticity in Ti Alloys at Low Temperatures: Ti-Nb Alloys. The extrapolated M_s curve of *Fig. 4-9* [EAS77][KOC77], although perhaps not to be taken literally, does, however, suggest the possibility of a $\beta \rightarrow \alpha''$ transformation. This and other indicators of lattice instability, such as the relatively low shear modulus for alloys of this class, Sect. 5.7, and the proven existence of diffuse ω phase, Sect. 4.3.3, is evidence in support of a possible stress-induced martensitic transformability for alloys whose temperature/composition states lie just above the dashed line; this includes the commercial superconducting alloy, Ti-50Nb, at low temperatures. If the transformation is reversible (i.e., pseudoelastic), no trace will remain upon the removal of stress and the return of the sample to room temperature [EAS75][PAS78]. Evidence for its occurrence may be *indirect*—such as the characteristic signature of the $\sigma(\epsilon)$ curve. It may also be *direct*—as in the *in situ* cold-stage TEM experiments of OBST et al. [OBS80], who noted what appeared to be a reversible transformation in reversibly strained Ti-45Nb.

Stress-strain behavior, thought to be characteristic of pseudoelasticity, has been detected in Ti-55Nb at 4.2 K by EASTON and KOCH [EAS75] [KOC77]. During the first loading cycle of a low-temperature tensile test, slip took place partially reversibly at a strain level of about 2.2%, but the sample subsequently executed closed hysteresis loops with no further accumulation of residual strain up to 100 cycles of deformation. The hysteretic deformation, which set in above the 0.4% strain level [EAS75] [PAS78], and which was accompanied by acoustic emission during both loading and unloading [KOC77][EAS75, EAS77], was interpreted as a manifestation of the pseudoelastic martensitic transformation. At higher strain levels, serrated yielding was encountered.

(b) Pseudoelasticity in Ti Alloys at Room Temperature: Ti-V and Ti-6Al-4V. Since strain cycling is needed to elicit the characteristic response, pseudoelasticity if present will manifest itself during the initial stages of cyclic fatigue testing in the form of: (i) an irregularity in an otherwise smooth stress-strain hysteresis loop, or (ii) an "anomalous Bauschinger effect". The Bauschinger effect itself (to be discussed in Sect. 5.14.6) is the premature yielding in compression (tension) of a material initially deformed in tension (compression). Clearly, a fully pseudoelastic material in which a removal of the tensile stress is sufficient to close the stress-strain hysteresis loop exhibits the greatest possible departure from normal Bauschinger behavior. Otherwise, partial pseudoelasticity will produce extreme premature reverse-stress yielding or cause undulations to develop in the reverse-stress-strain curve, either of which could be described as anomalous Bauschinger behavior.

CHAKRABORTTY *et al.* [CHA78[a]] have studied the cyclic stress-strain responses of a series of Ti-V(24, 28, 32, and 36 wt.%) alloys in the quenched and quenched + aged conditions. In only Ti-24V was an anomalous Bauschinger effect observed.* The mechanism in this case was deformation twinning under tension, followed by un-twinning when the stress was removed. A prerequisite was a low level of forward strain, during which small undamaged twins were produced and the twin-energy was conserved. The anomaly tended to go away at large strain amplitudes as stress-induced twins became damaged by mutual interaction and by interaction with dislocations, and energy became dissipated.

Pseudoelasticity gives rise to an exaggerated Bauschinger effect. On this evidence, acquired during both tensile [STE76] and torsional [GIL76] cyclic fatigue tests, some pseudoelastic character has been attributed to Ti-6Al-4V. The practical evidence is that Ti-6Al-4V exhibits "pseudoelastic fatigue failure". This is brought about by pseudoelastic slip, which in turn initiates crack formation at stress levels well below the macro-yield range. In the experiments of GILMORE *et al.* [GIL76], the slip appeared to be taking place at grain boundaries rather than within the grains.

5.14.5 The Shape-Memory Effect

The shape-memory effect, *Fig. 5-14(d)*, may be regarded as a combination of the previous two effects (*Figs. 5-14(b)* and *(c)*) in that after the initial stress-induced permanent transformation, the reverse transformation

---

*Quenched Ti-24V was β plus athermal ω. The other alloys were all-β and did not exhibit anomalous Bauschinger behavior.

is achieved by raising the temperature of the sample [DEL74]. The effect is usually associated with alloys exhibiting martensitic thermoelasticity (so-called "marmen" alloys), whose behavior and properties have been considered fully in reviews assembled by PERKINS et al. [PER75, PER75a]. Thermoelastic martensitic transformation and the shape-memory effect have also been reviewed by TONG and WAYMAN [TON75], with particular reference to the alloys Au-Cd, Ag-Cd, and Ti-Ni, and their use as the "working substances" in heat engines. Although the most famous shape-memory material is, of course, TiNi (so-called "NITINOL"), numerous alloys exhibit the effect which, as the following brief discussion indicates, is not necessarily confined to martensitic structures.

NISHIHARA and IGUCHI [NIS76] have detected what they referred to as shape-memory behavior in Ti-6Al-4V, which after superplastic deformation underwent partial length recovery as a result of $\alpha + \beta \rightarrow \beta$ transformation during stress-free heating. In experiments on Ti-Nb(22 at.%), which fully transforms to α'' on quenching through ~175°C, BAKER [BAK71] showed that 1% "permanent" strain could be almost fully (~90%) erased by heating to 184°C. Similar effects were noted for the same alloy in the $\omega + \beta$ state after aging for 2 and 4 min at 400°C, except that almost 100% recovery from $1\frac{1}{2}$% strain was achieved. Experiments have indicated that a shape-memory effect can occur either in an alloy undergoing a martensitic $\beta \rightarrow \alpha''$ transformation, or in a specimen whose initial structure is already fully martensitic, in which case the effect proceeds by way of a reorientation of the martensite plates [DEL74]. Since the $\beta \rightarrow \alpha''$ transformation is accompanied by considerable dilatation, its occurrence during a tensile test either isothermally (pseudoelastic effect) or adiabatically (shape-memory effect) will give rise to load drops and corresponding $\sigma(\epsilon)$ anomalies (serrated yielding), the forms of which will be characteristic both of the sample properties and the testing machine response times (see Sect. 5.14.1(a)).

5.14.6 The Bauschinger Effect

It could be argued that since the Bauschinger effect is generally observed during the stress-strain cycling of metals, it is not anomalous and should, therefore, have been considered in Part 2 of this chapter. On the other hand, it *is* anomalous in the sense that it involves stress-strain asymmetry explicable only in microstructural terms. In contrast to the repeated unidirectional application-and-release of stress, which leads to work-hardened elastic stress-strain reversal, cyclical loading/unloading/load-reversals lead to hysteretic stress-strain loops of the kind illustrated in *Fig. 5-14(e)*. Furthermore, it is generally found that the third quarter-cycle is accompanied by yielding at a prematurely low stress level as gauged by the parameters of the first. This is the Bauschinger effect, which may, therefore, be defined as reversible-stress-softening which permits premature yielding of a material upon stress reversal after an initial half-cycle involving plastic deformation (in either the compressive or tensile mode). It has been discussed by GITTUS [GIT75, p. 76] (in terms of dislocation effects) and TIMOSCHENKO [TIM56, p. 413] (in terms of an intergranular/intragranular mechanism). The effect is generally smaller the greater the strain during the initial half-cycle, and disappears completely with large amounts of strain hardening, Sect. 5.14.4(b) [TWE64, p. 74].

With regard to the *dislocation mechanism*, a "normal" Bauschinger effect would result when dislocation energy, stored during forward straining, adds to the applied stress during reverse loading. THEODORSKI and Koss [THE82] have discussed the Bauschinger effect in Ti_{61}-V_{38}-Si_1 and the technical alloys Ti-8Mo-8V-2Fe-3Al (Ti-8823) and Ti-3Al-8V-6Cr-4Mo-4Zr (β-C, Ti-38-6-44) from this standpoint.

TIMOSHENKO, whose approach is specific to polycrystalline metals, developed a mechanical model for the Bauschinger effect based on *intragranular slip and its anisotropy*. Expanding on this and related ideas, MARGOLIN et al. [MAR78] have suggested that the Bauschinger effect could arise as a result of residual stresses which accumulate in the presence of grains with different yield stresses. Assuming a tension/compression sequence, grains with lower yield stresses become placed in compression by the other grains, and thereby experience in addition to the applied stress an extra compressive stress. The theory was examined using Ti-15Mn as a test material [MAR78]. Presumably the same mechanism is also operative, but at a more macroscopic level, in certain bimetallic composites which by their very natures generally consist of pairs of metals with widely differing mechanical properties. An excellent example is the well-known Ti-50Nb/Cu superconductive multifilamentary composite, in which hundreds of fine filaments of Ti-50Nb, with a relatively high modulus, are imbedded in annealed OFHC Cu. The Cu yields at a much lower stress than the Ti-50Nb; consequently, it is possible for the Ti-Nb still to be stretching elastically long after the Cu has passed its yield point. As a result, on the return stroke of the testing machine, the Cu is placed in compression by a combination of the applied stress and the elasticity of the Ti-Nb [HEI74]. A somewhat similar effect should be encountered in two-phase alloys which could in this context be regarded as *in situ* metal-matrix composites. Bauschinger effect in two-phase Ti-Mn alloys has been considered by SALEH and MARGOLIN [SAL79a].

The above models of the Bauschinger effect have all invoked an additional hidden component of stress to augment the applied stress and bring about an "apparent" premature third-quadrant yielding. In an unstable lattice, stress-induced transformation or twinning is responsible for an "actual" premature softening and will result in an anomalous Bauschinger effect. As pointed out in Sect. 5.14.4, pseudoelastic materials provide examples of this. Anomalous Bauschinger effects are also exhibited by alloys such as Ti-6Al-4V, which after being strained in tension (from the as-received condition) to a 1% permanent set, suffer a 40% loss of compressive yield stress [Woo72, p. 1-4:72-9]. As mentioned above, the anomalous Bauschinger effect in this alloy is thought to be caused by yielding at the grain boundaries [Gil76].

Deformation

Probably the most common laboratory-scale deformation process is that experienced by samples in the final stages of uniaxial compressive or tensile testing. The most interesting properties to be encountered under such conditions are the so-called "anomalous tensile properties" (e.g., serrated yielding, pseudoelasticity, etc.), the subjects of the latter part of the preceding chapter.

The manner in which a metal deforms after its yield strength has been exceeded by the applied stress depends on many factors. Parameters controlling the deformation process include the alloy's *composition*, its *class* (i.e., whether α, $\alpha + \beta$, or β), its *condition* (i.e., whether quenched — e.g., β_{tr}, $\alpha + \beta$-annealed, low-temperature aged, etc.), and the *rate* and *temperature* at which the deformation is carried out. Some *observables* or results of the deformation process include the *anomalous* stress-strain behavior alluded to above and discussed in Chapter 5, *phase transformation* under stress (i.e., transformation-assisted deformation), and *texturization* (i.e., the development of preferential crystal orientation or the formation of deformation cells or subbands in response to heavy cold work).

Some examples of these processes and effects are offered below under the headings: (i) ductility at low temperatures, (ii) deformation at elevated temperatures (forming), (iii) transformation-assisted plasticity, (iv) deformation textures, and (v) deformation microstructures.

6.1 Low-Temperature Ductilities of Some Representative Technical Ti-Base Alloys

As mentioned in Chapter 1, unalloyed Ti, the α alloy Ti-5Al-2.5Sn, and the near-α $\alpha + \beta$ alloys Ti-6Al-4V and Ti-8Al-1Mo-1V have properties which are suitable for a wide range of cryogenic applications, while β alloys such as Ti-13V-11Cr-3Al have a strong tendency to embrittle on cooling to cryogenic temperatures. The *low-temperature* mechanical and physical properties of these alloys have been specified in SALMON's *Low Temperature Data Handbook* [SAL79].

6.1.1 Unalloyed Ti

All commercial grades of unalloyed Ti exhibit moderately good ductility at temperatures down to ~20 K. Their elongations to fracture (δ_B, %) actually increase as the temperature is decreased from 300 K, and pass through broad maxima ($\delta_B = 40 \sim 50\%$) at about 77 K before descending rapidly as the temperature approaches 4.2 K. In some samples, δ_B becomes negligibly small at liquid-He temperatures. Cold rolling increases the yield and ultimate strengths, but at the expense of ductility, as usual. The effects of interstitial elements on the strength of Ti have been considered in great detail by CONRAD and co-workers, Sect. 5.8.1(b). In a long series of papers it has been pointed out that the solutes C, N, and O, which bond in a covalent-like manner to the surrounding Ti atoms, have pronounced influences on the strength of otherwise unalloyed Ti at temperatures below about 0.5 T_m.

6.1.2 Ti-5Al-2.5Sn

The ductility of extra-low-interstitial (ELI) grade Ti-5Al-2.5Sn (with O, 1200; C, 800; N, 500 max. ppm by wt.) is fairly independent of temperature between room temperature and 20 K, δ_B remaining at about $16 \pm 1\%$ throughout that range. The ductility of the normal-interstitial grade (O, 2000; C, 1500; N, 700 max. ppm by wt.) is considerably lower; in fact, δ_B decreases monotonically below room temperature, dropping to 12% at 77 K and to only 5% at 20 K.

6.1.3 Ti-6Al-4V

The ductility of annealed Ti-6Al-4V is fairly temperature independent between room temperature and 77 K. Below that, it decreases rapidly as the temperature continues to lower toward 20 K, *Fig. 6-1*. The ductility of

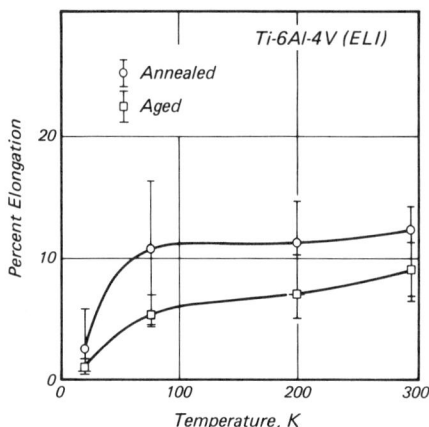

Fig. 6-1. Ductility of the extra-low-interstitial (ELI) grade of Ti-6Al-4V as a function of temperature in the cryogenic-to-room-temperature range [SAL79].

the annealed alloy is twice as great as that of the solution-treated-and-aged (STA) material; e.g., $\delta_{B,77\,K} = 11.4\%$ as compared to 4.9% (at normal-interstitial levels). Reducing the interstitial content influences the tensile properties only marginally, but improves the fracture toughness by 130% at room temperature and 40% at 20 K. For this reason the ELI grade of the alloy is recommended for cryogenic service.

6.1.4 Ti-8Al-1Mo-1V

The near-α $\alpha + \beta$ alloy Ti-8Al-Mo-1V, although originally developed for high-temperature applications, can be used reliably down to moderate sub-ambient temperatures in either the single-annealed (SA) or duplex-annealed (DA) conditions.* The room-temperature ductilities of SA and DA alloys are similar ($\delta_B \cong 15\%$), but upon cooling, that of the SA alloy decreases, while that of the DA increases before passing through a maximum ($\delta_B \cong 22\%$) at about 77 K and dropping to low values at 20 K ($\delta_B \cong 1\%$).

6.1.5 Ti-13V-11Cr-3Al

As a β-Ti alloy, Ti-13V-11Cr-3Al would be expected to possess poor low-temperature ductility. Indeed it does, the elongation-at-fracture of the STA material becoming insignificantly small below about 100 K. At 77 K, $\delta_B = 0.2\%$. Some improvement results if the aging stage of the STA heat treatment, (20 ~ 100h)/(430 ~ 500°C), is omitted, in which case $\delta_{B,77\,K}$ then becomes about 2%.

6.2 Deformation at Elevated Temperatures

As an introduction to the topic of elevated-temperature deformation, the forming requirements of the representative technical alloys of the previous section will be briefly reviewed.

(a) **Unalloyed Ti (α/β Transition Temperature, 883°C).** Commercial grades of Ti are usually hot-formed at 870°C, just below the allotropic transformation temperature. If cold deformed, the material will exhibit "spring-back" and will also require some annealing.

(b) **Ti-5Al-2.5Sn ($\alpha/(\alpha + \beta)$ Transition Temperature, ~1025°C).** The formability of Ti-5Al-2.5Sn is inferior to that of commercial unalloyed Ti. Forming operations are usually carried out at 200 ~ 650°C, care being taken to minimize the time of exposure to temperatures above 540°C.

(c) **Ti-6Al-4V (($\alpha + \beta$)/β Transition Temperature, ~995°C).** Ti-6Al-4V is difficult to form even after annealing—usually for ($\frac{1}{2}$-4h)/(700-820°C). As

---

Single-annealed: "mill annealed" (8h/790°C) and furnace cooled. *Duplex-annealed*: mill annealed plus 15min/790°C plus air cooled.

described by Wood [Woo72, pp. 1-4:72-3, 72-6], primary fabrication operations are usually performed at temperatures above the β transus, while the finishing of mill products usually takes place in the temperature range 870 ~ 980°C — i.e., high in the $\alpha + \beta$ field. Other forging practices, and the properties that result from them, are discussed by Wood [Woo72].

(d) **Ti-8Al-1Mo-1V (($\alpha + \beta$)/β Transition Temperature, 1040°C).** In the deformation metalworking of Ti-8Al-1Mo-1V, temperatures in the β field are generally avoided. It has been found preferable to perform metal-working operations high in the $\alpha + \beta$ field. Sheet metalworking (second-ary fabrication) is generally conducted within the temperature range 650 ~ 800°C, although sheet forming is not impossible at temperatures as low as room temperature.

(e) **Ti-13V-11Cr-3Al (Metastable β, β-Transus Temperature, 650 ~ 700°C).** Ti-13V-11Cr-3Al is normally fabricated to flat-rolled products at tempera-tures in the β-phase field [Woo72, p. 1-9:72-3]. As for secondary fabrica-tion, because of the alloy's excellent bend ductility, any sheet-forming operation that is applicable to unalloyed Ti is suitable for annealed Ti-13V-11Cr-3Al.

6.2.1 β Forging of Ti-Base Alloys

(a) **α-Forging Processes [Ham78].** β forging may be carried out isothermally, with the billet and dies initially at the same temperature above the β transus. Otherwise, if only the billet has been heated, die-chilling as the forging takes place may cool the work-piece to tem-peratures within the $\alpha + \beta$ field. Provided the first 25% of the reduction takes place within the β field, it is still permissible to refer to the operation as "β forging". If the initial temperature is sufficiently high in the β field, the forging can be accomplished as a single step. A second advantage of β forging is that at the high temperature at which it takes place the Si (in Si-bearing alloys such as Ti-6242(Si)) can be retained in solid solution during the operation, thereby leading to a product with improved creep strength. On the other hand, the high temperatures associated with the β forging of α and many $\alpha + \beta$ alloys may result in: (i) a large β grain size, particularly if the alloy is allowed to anneal in the β regime before cooling below the transus, and (ii) a coarse Widmanstätten structure on cooling to room temperature. Such an "aligned-α", or locally textured, structure has poor low-cycle fatigue properties. This disadvan-tage can be reduced if the aligned-α is replaced by the "basket-weave" structure by increasing the cooling rate from the β field, or by refining the prior-β grain size through final forging high in the $\alpha + \beta$ field.

(b) **Superplasticity in β Forging.** Isothermal β forging of β alloys can take advantage of the superplastic properties which have been exhibited by some of them. GRIFFITHS and HAMMOND [GRI73] showed that the alloys Ti-8Mn, Ti-15Mo, and Ti-13Cr-11V-3Al exhibited superplasticity when deformed at low strain rates at temperatures of about 0.6 T_m. Elongations of from 150% to 450% were observed, depending on the numbers of grains in the specimen cross-sections. With grain sizes of several hundreds of μm, the sample could not have been exhibiting the normal kind of fine-grain (<10 μm) superplasticity. The *subgrains* (or deformation cells) were, however, fine and of the above-mentioned size. The flow was interpreted as taking place by way of a "diffusion-creep" mechanism. In this, the vacancy source/sink property of the subgrain boundaries was supposed to combine with the anomalously high diffusion rates inherent in β-Ti alloys to permit the extensive low-strain-rate moderate-temperature deformation which characterizes superplasticity [GRI73][HAM78].

6.2.2 $\alpha + \beta$ Forging of Ti-Base Alloys

(a) $\alpha + \beta$ **Forging Process [HAM78].** Although during non-isothermal forging, as defined above, the billet may spend a significant fraction of its deformation time in the $\alpha + \beta$ field, in order to achieve a *uniform* $\alpha + \beta$ structure the entire operation must be carried out in that field. The processing temperature is then, of course, upper limited by the β transus and lower bounded by press and materials constraints. Such a tight restriction on processing temperature range may require preheating the tooling to the initial billet temperature (i.e., the use of isothermal forging).

(b) **Superplasticity in $\alpha + \beta$ Forging.** As indicated above, isothermal forging at low strain rates is conducive to superplastic deformation. In $\alpha + \beta$ alloys, superplasticity generally takes place via the fine-grain-size mechanisms. This is achieved in $\alpha + \beta$ alloys through heavy hot working in the $\alpha + \beta$ phase field. Continuous working and recrystallization produces a fine granularity, which is then stabilized by the presence of a coarse dispersed phase. Obviously, and as HAMMOND and NUTTING have pointed out [HAM78], there may not be a very sharp distinction between this kind of superplasticity and isothermal forging. The isothermal closed-die approach has been successfully applied to the superplastic forging of Ti-6Al-6V-2Sn and Ti-6Al-4V in the temperature range 900-950°C [FIX73].

6.2.3 Thermomechanical Processing

The properties of $\alpha + \beta$ Ti-base alloys are strongly dependent on the microstructure. CHEN and colleagues, for example, have devoted considerable attention to the relationship between the microstructures and the mechanical properties of Ti-6Al-2Sn-4Zr-2Mo~0.1Si forgings [CHE80]. Sects. 6.2.1 and 6.2.2 have briefly indicated the characteristics of β and $\alpha + \beta$ forging, both isothermal and non-isothermal. Factors which must be taken into consideration in the design of forging operations in general are: (i) the starting microstructure, (ii) the start and finish temperatures, (iii) the extent of the deformation, and (iv) the rate at which the deformation takes place [GEG80]. In achieving the desired final microstructure, hence mechanical properties, two more thermal variables are available for control and adjustment: (i) the cooling rate from the final working operation, and (ii) the final heat treatment. The entire sequence of operations, known as "thermomechanical processing" [WIL82[b]], is fully discussed in other volumes of this monograph series.

6.3 Transformation-Assisted Plasticity

The deformation of alloys can be facilitated if thermally induced or stress-induced transformations of one kind or another take place during the time that stress is being applied to the sample. Either high-temperature or low-temperature transformations are eligible for consideration in this section, which draws implicitly on the contents of Sect. 4.2 (stress-induced martensitic transformation and twinning) and Sect. 5.14 (low-temperature and high-temperature serrated yielding) as well as other sources.

6.3.1 β Alloys

Stress-assisted transformation permits some β alloys to achieve a degree of low-temperature ductility otherwise unexpected in a bcc structure. For example, Ti~50Nb, whose deformation at low temperatures has already been discussed in Sects. 4.2.5(c) and 5.14(a), when tested to failure at 4.2 K, reveals the finely dimpled fracture surfaces characteristic of microscopic ductile fracture, *Fig. 6-2*.

Numerous binary β-Ti alloys have been shown to deform via martensitic transformation [ZWI74, p. 175] or twinning, depending on the solute concentration. In a typical basic study, that by OKA *et al.* [OKA80] of Ti-Mo (see Sect. 4.2.5(b)), it was shown that at room temperature the stress-induced transformation product was martensitic provided the Mo concentration lay between about 9 and 11 wt.%, but took the form of $\{332\}_\beta$ twins when it was within the range 11 ~ 15.5 wt.%. In a related study, HIDA *et al.* [HID80] investigated the relationship between the thermal instability of metastable quenched β-Ti-Mo and its plastic properties, observing again that the remarkable ductility exhibited by Ti-14Mo correlated with the formation of $\{332\}_\beta$ twins. Secondly, they noted that the measured linear work-hardening rate could be correlated with a continuous formation of new twins, these being the source of an increasing density of ω phase, a known hardener. As already mentioned in Sect. 4.3.3(b),

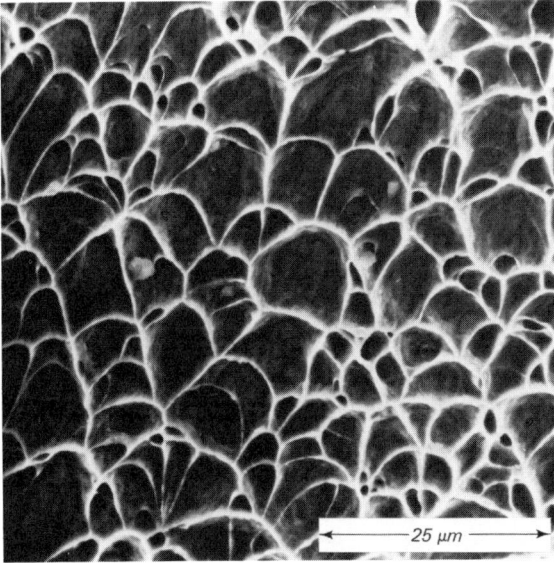

Fig. 6-2. Honeycombed or dimpled shear-fracture surface in Ti-50Nb (90% cold deformed) [ALB76]. Copyright © 1976, *Zeitschrift für Metallkunde*, reprinted with permission.

the β alloy Ti-10V-2Fe-3Al will undergo stress-assisted transformation at *room temperature*, which lies part-way between M_d and M_s [DUE80^a]. The β-solution-treated alloy supports a fine dispersion of athermal ω precipitation after quenching. Upon stressing at room temperature, the matrix distorts to α'' martensite, leaving the ω precipitates intact—a phenomenon which imparts some ductility to an alloy that might otherwise be embrittled by ω phase.

6.3.2 α and $\alpha + \beta$ Metals and Alloys at Elevated Temperatures

Since, as has been generally noted, a metal's resistance to plastic deformation decreases during a phase change, phase transformation can be brought in to assist in elevated-temperature deformation. GEGEL, NADIV, and RAJ, in a study of the dynamics of flow and fracture during the isothermal forging of Ti-6Al-2Sn-4Zr-2Mo-0.1Si, have noted that high-strain-rate deformation at temperatures some 50-150°C below the $\beta/(\alpha + \beta)$ transus would induce the formation of microcracks [GEG80]. This crack formation could, however, be suppressed by reducing the strain rate below some critical value in order, it was thought, to permit the operation of a phase-transformation type of stress-relaxation mechanism. Another example is to be found in the work of KOT *et al.* with unalloyed Ti and Ti-6Al-4V

[Koτ70]. In a typical experiment, the sample was subjected to fixed loading while its temperature was cycled through the β transus (650-925°C for Ti, and 760-980°C for Ti-6Al-4V). Each time the alloy underwent an $\alpha \rightarrow \beta$ transformation, quasi-viscous flow took place, accompanied by a large increment of plastic deformation. Upon repeating such cycling, step-wise elongations of Ti and Ti-6Al-4V in excess of 300% are possible and have been observed. In an interesting application of the principle, Koτ *et al.* [Koτ70] employed a traveling induction coil as a kind of "thermal die". Looped around a rod of Ti or Ti-6Al-4V, maintained under tension, it was capable of administering area reductions in excess of 50% during each of its passes along the rod.

6.4 Deformation Textures and Microstructures in α- and β-Ti Alloys

The two metalworking processes that administer the greatest amounts of deformation from ingot to finished product are *sheet rolling* and *wire drawing*. As a result of the heavy unidirectional deformation which these processes impart, the resulting sheets or wires acquire certain directional properties either at the atomic level (as gauged by the results of x-ray diffractometry) or at the microstructural level (as visualized by optical and electron microscopy).

6.4.1 Texture

Texture refers to the tendency for the principal crystallographic directions in adjacent grains of a polycrystalline material to become aligned, or to assume a "preferred orientation", *Fig. 6-3*. Texture is represented by a two-dimensional map, or pole figure, which (as the projection of a

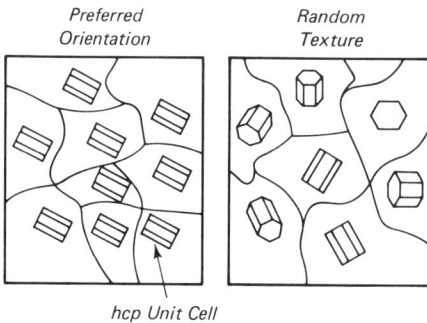

Preferred Orientation   *Random Texture*

hcp Unit Cell

Fig. 6-3. Schematic representation of preferred and random textures in a polycrystalline hcp metal.

hemispherical surface onto a flat one) quantifies the angular clustering of selected crystallographic directions or poles. Since the existence of texture is amplified if the individual crystallites are themselves highly anisotropic, Ti-alloy texture has usually been studied in unalloyed Ti, dilute α-phase alloys (particularly Ti-Al), or on the near-α α + β alloy Ti-6Al-4V. The β alloys have not, of course, been neglected; in fact, binary alloys of Ti with V, Mn, Nb, and Mo [MAR60] and alloys such as Ti-16V-2.5Al [LAR74] have been extensively studied. The subject of texture has been reviewed numerous times: the early work, by authors such as JAFFEE [JAF58, pp. 100-101] and MARGOLIN [MAR60, pp. 269-271]; the more recent results, by LARSON and ZARKADES [LAR74] and ZWICKER [ZWI74].

6.4.2 Deformation Microstructures

Whether or not a material becomes crystallographically textured, heavy deformation will produce aligned arrays of dislocation-cell walls, *Fig. 6-4*. Our present understanding of the effects of heavy deformation on the microstructures of metals can be traced back to an important series of papers by HIRSCH and co-workers (Refs. (53)(54)(55) of [NAR66]), the essential results of which, as they apply to bcc metals, having been elegantly summarized by NARLIKAR and DEW-HUGHES [NAR66]. With the aid of a "microbeam" Laue back-reflection technique, in which the x-ray beam could be collimated through a capillary <20 μm in diameter, HIRSCH and co-workers were able to recognize the presence of crystallites with diameters as small as 1 μm. The importance of this work lay in its being able to show quantitatively that: (i) heavy deformation did not lead to a uniformly disordered structure as had earlier been thought but rather, in

Fig. 6-4. Longitudinal section of a cold-swaged (97.2%) Ti-50Nb wire [LOH71]. Micrograph courtesy of U. Zwicker (Universität Erlangen-Nürnberg).

their words, to a "foam"-like structure consisting of particles of low dislocation density imbedded in a continuous three-dimensional net of highly dislocated material; and (ii) the resulting structure, instead of being the product of grain disintegration (and consequently true grain refinement), was in fact generated by dislocation motion within the grains such that the resulting dislocation network formed low-angle subgrain boundaries. In a related study, EMBURY *et al.* [EMB66] investigated and intercompared the microstructures of several steels and a sample of 99.99% Cu in response to wire drawing at room temperature to area reductions of up to 99%. Again, it was noted that all the samples developed fibrous or cellular structures, the cell walls acting as dislocation barriers in a manner analogous to the function of grain boundaries in this regard. A much greater degree of cell refinement was noted in the Fe alloys than in the Cu, due it was claimed to a difference between the rates of dynamic recovery in the two cases. It was thought that the presence of interstitial impurities in the former was responsible for the greater stability of the substructural boundaries once they were produced by the wire-drawing process. As a consequence of the wire drawing and the resulting fine elongated cellular structure, the steels became significantly strain hardened. For each steel the flow stress, σ_f, increased linearly with $1/\sqrt{d}$, where d is the cell diameter. To a first approximation, σ_f (kg mm^{-2}) = $40/\sqrt{d}$, if d is expressed in μm. In the English scientific literature these subgrains have also been referred to as "cells" or "sub-cells" and the dislocation network defining them as "cell/sub-cell walls/boundaries"; the German* and more recent English literatures seem to prefer the use of "subbands" and "subband boundaries" or "walls" to describe the same features. The presence of individual dislocation-free cells was disclosed by the appearance of spots on the Laue microdiffraction rings (Refs. (53) and (54) of [NAR66]). Later investigators, studying heavily cold-drawn [ARN74] or cold-swaged [LOH71] wire using both microbeam and conventional diffraction methods, have interpreted the spot patterns as being indicative of $\langle 110 \rangle$ texture in the direction of the wire axis.

6.5 Deformation-Induced Textures in Ti-Base Alloys

During deformation metalworking, or as a result of recrystallization, grain growth, or phase transformation, the grains (crystallites) of polycrystalline metals may develop preferred orientations or texture. The topic

---

*According to HILLMANN [HIL73], "deformation bands" are 10^{-4} and 10^{-5} cm in width, while "subbands" formed as a result of still stronger deformation are 10^{-5} to 2×10^{-6} cm wide.

has been subjected to a comprehensive general review by DILLAMORE and ROBERTS [DIL65] (407 references). With regard to Ti and its alloys, reviews have been offered by JAFFEE [JAF58], MARGOLIN [MAR69], LARSON and ZARKADES [LAR74], and ZWICKER [ZWI74], authors who have focused attention primarily on the results of cold rolling which, of course, is capable of developing strong, reproducible, preferred orientations.* The development of texture during the *forging* of Ti-alloy parts has been considered by ROMERO [ROM71].

Texture is determined by x-ray diffraction and defined usually by means of a single-quadrant stereographic representation of the clustering of specified crystallographic orientations (the "pole figure"). Texture in hcp polycrystals is often described in terms of the distribution of *c*-axis or [0001] directions. The [0001] direction in the hcp crystal is referred to as the "basal pole"; it is also referred to as the "(0002) pole", the implication here being that the basal planes of an hcp crystal are responsible for (0002) Bragg reflections. In short, the terminologies "[0001]" and "(0002)" that occasionally appear together, and even interchangeably, in the literature of hcp texture refer to *crystalline direction* and *experimental method*, respectively. A pair of schematic basal-pole quadrant-diagrams are given in *Fig. 6-5.*

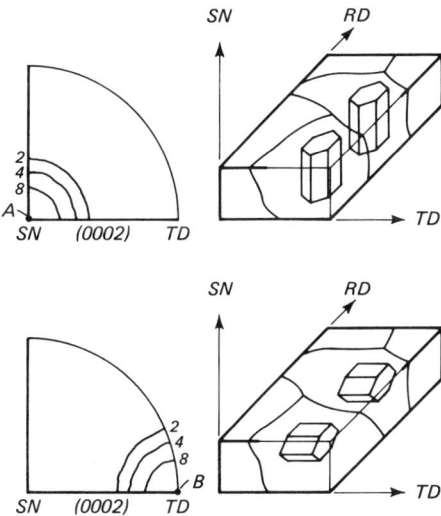

Fig. 6-5. Schematic representation of basal pole figures for textured hcp metals. *Upper diagram*: crystallites aligned normal to the rolling direction; point *A* would represent perfect alignment of all the crystallites. *Lower diagram*: crystallites aligned transverse to the rolling direction; point *B* represents perfect alignment in this case [LAR74].

---

*Sheet textures can be defined by specifying the Miller indices of a plane parallel to the rolling plane (*hkl*) and a direction parallel to the rolling direction [*uvw*], see [LAR74].

6.5.1 Unalloyed Ti

The cold-rolled texture of unalloyed Ti has been determined many times either as a study in itself (e.g., [KEE 56]) or as part of an investigation into the influence of alloying on textural change (e.g., [THO 73]). As with most low-c/a (<1.63) metals, the (0002) or basal poles are concentrated in regions $\pm 30°$ to $\pm 40°$ in the transverse direction away from the sheet normal, *Fig. 6-6(a)*. All authors agree that such a texture results from a competition between $\{0001\}\langle 11\bar{2}0\rangle$ slip, which rotates the basal poles toward the sheet normal, and $\{11\bar{2}2\}$ twinning, which tends to rotate them into the transverse direction.

The annealing of high-purity Ti at various temperatures below the transformation temperature (883°C) results in a sharpening and slight "rotation" of the texture; i.e., it leaves the (0002) pole figure essentially unchanged, but rotates the (10$\bar{1}$0) pole figure about the sheet normal. Heating through the α/β transition and back again does not erase the texture in either phase, presumably because the Burgers relationship, $\{110\}_\beta \| (0001)_\alpha$, holds during both the positive and negative temperature excursions [JAF 58, p. 101].

6.5.2 α-Phase Binary Alloys

Textures in Ti-Zr(0.03-9.04 wt.%), Ti-Sn(0.01-4.00 wt.%), and Ti-Al(0.01-3.92 wt.%) alloys have been measured by LARSON *et al.* [LAR 71]. Ti-Al has, of course, been extensively studied—first by McHARGUE *et al.* [MCH 53][SPA 57] and subsequently by LARSON *et al.* (just mentioned) and by THORNBURG [THO 73]. The early work showed that in the presence of sufficient Al, certainly with 3.8 wt.% of it [MCH 53], cold rolling produced an almost "ideal" basal texture (i.e., one in which the basal poles were clustered about the sheet normal) but left open the question as to whether the transition to this, from the split texture of unalloyed Ti, took place suddenly at some critical Al concentration or was a smooth, continuous function of it.* To answer this question, THORNBURG [THO 73] prepared for measurement alloys of Ti with 0.25, 0.5, 1.0, 1.5, 2.0, 3.0, and 4.0 wt.% Al and measured their textures after 20, 40, 60, 80, 90, and 95% area reduction by cold rolling. They discovered that not only did the texture maintain a fairly constant degree of splitting up to an Al concentration of 2 wt.%, but that it transformed suddenly to basal at that concentration and remained that way as the Al concentration continued to

---

*[MCH 53] and [SPA 57] between them treated Ti-Al(0.07, 0.47, 1.05, 1.43, and 3.8 wt.%).

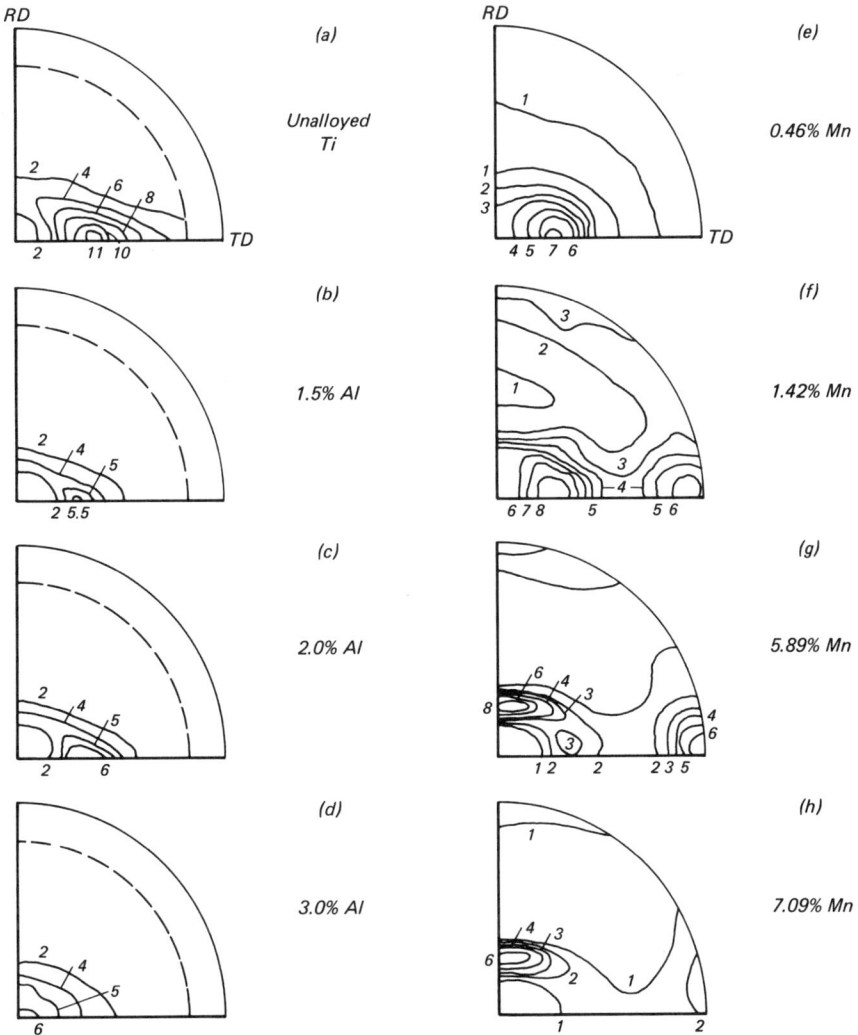

Fig. 6-6. Textures of cold-rolled Ti-Al and Ti-Mn alloys. Sample conditions: *Ti-Al series*: hot rolled 50°C above $\beta/(\alpha+\beta)$, then hot rolled 50°C below $\alpha/(\alpha+\beta)$ and annealed there for 1 h; cold rolled 95% [THO73]. *Ti-Mn series*: forged at 840°C, hot rolled at 815°C, then cold rolled 84% with intermediate anneals at 730°C [LAR71].

increase. They also noted that, with the exception of the threshold-concentration alloy (2 wt.% Al), the texture had reached a stable saturated condition by 60% reduction in area. A series of (0002) pole figures taken at 95% reduction by cold rolling is given in *Figs. 6-6(b)* through *(d)*.

6.5.3 α + β-Phase Binary Ti-TM Alloys

The textures of binary Ti-TM alloys, hot-worked in the $\alpha + \beta$ field and subsequently cold-rolled 84% with intermediate annealing, were investigated by LARSON et al. [LAR 71]. The entire Ti-(3d)TM series (TM = V to Ni) was measured, as were representatives of the two Ti-(4d)TM systems: Ti-Nb and Ti-Mo. Solute concentration ranges were chosen so as to provide maximum β-phase fractions of some 15 ~ 40%. Among all of the systems some striking similarities were to be seen in the manner in which the (0002) texture varied with solute concentration. A typical series of results, that for Ti-Mn(0.46, 1.42, 5.89, and 7.09 wt.%), is given in *Figs. 6-6(e)* through *(h)*. At low solute concentrations, the split basal-pole texture remained practically unchanged with increasing concentration except that, in some cases, it was slightly perturbed by the appearance of a small fraction of basal poles in the transverse direction.* Then after what seemed to be a critical solute concentration (corresponding to the presence of 16 ~ 20% β phase in the $\alpha + \beta$ alloys), the split basal texture shifted to the rolling direction. The texture of cold-rolled Ti-Cu was not a member of the above class: according to LARSON et al. [LAR 71, LAR 74], the addition of 0.55 wt.% Cu to Ti resulted in an almost "ideal" (0001)[10$\bar{1}$0]-type texture.

6.6 Deformation Microstructures in β-Ti Alloys

6.6.1 Rolling-Induced Microstructures

Relatively little attention has been given to the study of rolling-induced deformation microstructures in β-Ti alloys. Such information as is available has been reviewed by ZWICKER [ZWI 74, p. 134]. In an early study of type-II superconductivity, HAKE, LESLIE, and RHODES [HAK 63] noted that the cold rolling of Ti-Nb(60 at.%) to a thickness reduction ratio of 24:1 resulted in the formation of a pronounced laminar cell structure when viewed either transversely or longitudinally within the rolling plane. The cells, which appeared under optical magnifications to be continuous across the width of the sample, were about 2 μm thick. TEM observations of the in-plane structure of a sample of Ti-Nb(75 at.%), cold rolled to more than 90%, revealed a well-defined network of cells about 0.4 μm in size [NAR 66]. In order to completely evaluate the cell structure of highly anisotropic cold-rolled Ti-Nb, BAKER and TAYLOR [BAK 67] carried out a TEM study of 97% cold rolled Ti-45Nb, taking observations both normal

---

*For further details regarding this peak, a (11$\bar{2}$0)[10$\bar{1}$0] texture, and the (10$\bar{1}$0) pole figures, the original literature [LAR 71] should be consulted.

to the rolling plane and parallel to it along the rolling direction. The diameters of the in-plane cell cross-sections, which were fairly equiaxed, were 0.25 μm or larger, in good agreement with the earlier result [NAR 66]. But as expected, the end-view was one of severely flattened cells, about 0.1 μm thick and 2.5 μm or more in width.

6.6.2 Wire Flattening

Wire flattening consists of the application of light rolling to wire already deformed by drawing. In a cold-flattening study by BAKER and TAYLOR [BAK 67], a wire, cold drawn 99.87% to 0.033 in$^\phi$, was lightly cold rolled (4.7:1) to a thickness of 0.007 in. In consequence, an initial structure of parallel fibers ~0.1 μm in diameter underwent a coarsening when viewed normal to the rolling plane, and dislocation cells appeared within some of the fibers. It was expected that further cold reduction of this type would transform the structure to that observed in rolled material, which has been shown to consist of equiaxed cells in the rolling plane and a fibrous structure at right angles to it. Indeed in a similar experiment, in which heavily cold-drawn (99.994%) Ti-50Nb wire was reduced from a diameter of 0.25 mm to a thickness of 0.1 mm by cold rolling, PFEIFFER and HILLMANN [PFE 68] noted that whereas the in-plane structure was smeared out, especially along the edges, in the section at right angles to the rolling plane the initial fibrous structure of the wire had scarcely been disturbed.

6.6.3 Wire Drawing

As a uniaxial deformation process, wire drawing produces uniform deformation across a plane at right angles to the wire axis. Numerous photomicrographs are available in the literature showing an equiaxed cell structure in cross-section and the usual elongated bands in the axial direction. One such pair of micrographs, that for Ti-58Nb cold drawn to an area reduction ratio of 5×10^4:1 (99.998%) [NEA 71], *Fig. 6-7* and *Fig. 6-8*, depicts the microstructure of heavily cold-drawn wire as bundles of "pencil-shaped sub-cells" [NEA 71] or "fibers" [BAK 67] running parallel to the wire axis.

In studies of cold-drawn texture in β-Ti-Nb, since sample smallness precluded the use of normal x-ray pole-figure determination methods, ARNDT and EBELING [ARN 74][WIL 75[a]] employed the electron beam (20 μm$^\phi$) microdiffraction technique referred to above to view thinned cross-sections of the wire. Although heat-treated wire, in which some cell growth had taken place, yielded single-crystal diffraction patterns, results interpretable in terms of texture could be obtained from wires with narrow and well-defined straight subbands. In such cases, the diffraction pattern of numerous sharp spots representative of a superposition of many crystal

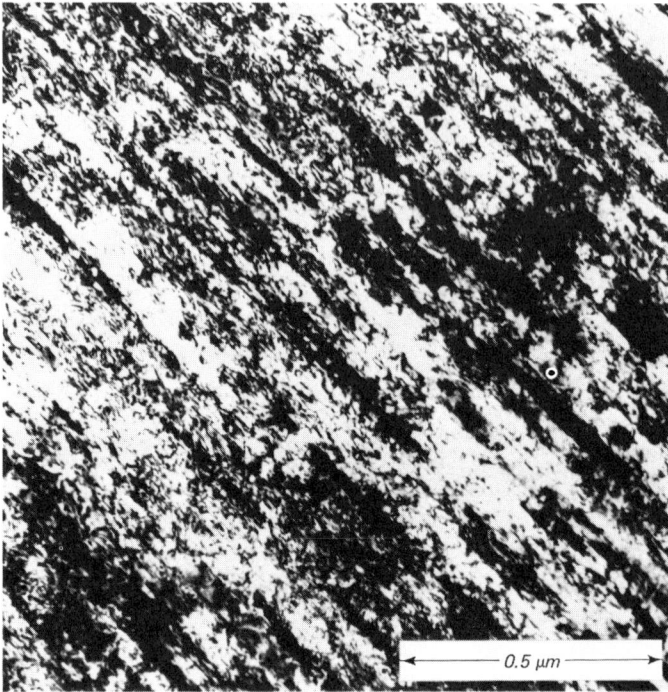

Fig. 6-7. Longitudinal section of a cold-drawn (ARR, 5×10^4:1) Ti-58Nb wire [NEA71]. Micrograph courtesy of D.F. Neal (Imperial Metal Industries); copyright © 1971, Pergamon Press, reprinted with permission.

orientations suggested a strong $\langle 110 \rangle$ texture in the axial direction.* The results of other measurements of Ti-Nb [HIL81] as well as of Ti-50Nb-1Cu and Ti-Mo [ZWI74, p. 136] were in agreement with this. Cu, which is fcc, was reported to develop a $\langle 100 \rangle$ texture after plastic deformation [REE77].

Although the subbands in as-deformed wire are far from regular, it has been possible to quantify the decrease in fiber diameter which accompanies an increase in the degree of cold work as gauged by the area-reduction ratio, ARR. For example, PFEIFFER and HILLMANN [PFE68] showed that a subband density (i.e., number of cells per cm^2 in the transverse section) of 1.3×10^{10} cm^{-2} in 99.78% cold-drawn ($32 \rightarrow 1.5$ mm$^\phi$) Ti-50Nb increased to 4.9×10^{11} cm^{-2} after 99.993% ($32 \rightarrow 0.25$ mm$^\phi$) reduction. The cell diameters naturally decreased from 0.09 to 0.01 μm. Earlier results by BAKER and TAYLOR [BAK67] for 99.87% cold-drawn Ti-55Nb

---

*A fiber texture can be defined in terms of the crystallographic direction which points along the axis of the cylindrical product; the other crystallographic directions are disposed randomly about this axis [LAR74].

0.5 μm

Fig. 6-8. Transverse section of the wire of *Fig. 6-7* [NEA71]. Micrograph courtesy of D.F. Neal (Imperial Metal Industries); copyright © 1971, Pergamon Press, reprinted with permission.

with a cell diameter of about 0.1 μm were in good accord with this, as were later data by NEAL *et al.* for Ti-58Nb [NEA71] and by ARNDT and EBELING [ARN74] [WIL75ᵃ] again for Ti-50Nb. A survey of these results shows that, in spite of the variation in Ti-Nb alloy composition, a remarkably good picture of the inverse correlation between fiber diameter and degree of cold work can be assembled. Data from these sources, and others [WES80], when plotted semilogarithmically as in *Fig. 6-9*, indicate that subband diameter in the range 100-1000 Å is negatively correlated with $\log_{10}$ ARR. A least-squares fit to the data yields a semiquantitative relationship (correlation, -65%) between subband diameter, d (which is never much less than about 450 Å), and ARR of the form:

$$d = \frac{(5.49 \pm 0.10) - \log_{10}\text{ARR}}{(320 \pm 26) \times 10^{-5}} \ (\text{Å}) \ , \tag{6-1}$$

in which the " $+$ " signs are to be taken together when the 140 Å datum point is *excluded* from the slope analysis; the " $-$ " signs, when it is

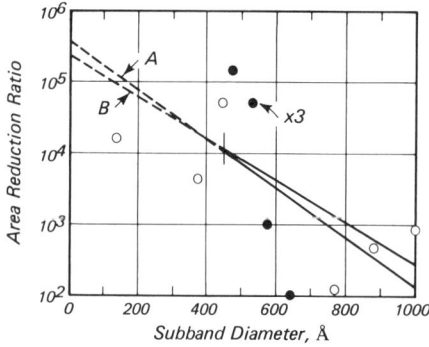

Fig. 6-9. Subband diameter as a function of area-reduction ratio (ARR) by cold drawing. The lines are least-squares fits to the data represented by the open (as-drawn) and closed (cold-drawn-plus-aged 385°C) circles. Lines *A* and *B* correspond to the omission and inclusion, respectively, of the point (140 Å, 1.6×10^4).

included. Notice that the deformations imposed in order to develop these fine subband structures are orders of magnitude greater than those needed simply to texturize the sample (cf. *Fig. 6-6*).

7

Aging

Part 1: Microstructural Phenomenology

The terms "aging" and "tempering" refer to moderate-temperature heat treatments during which diffusion-controlled metallurgical processes take place within macroscopic periods of time, measured in minutes and hours. The aging of metastable alloys is accompanied by precipitation as they proceed, generally by means of a nucleation-and-growth mechanism, toward thermodynamic equilibrium. An excellent treatment of the thermodynamics of such processes has been produced by HARDY and HEAL [HAR 56]. Aging can also assist in the recovery of a deformed structure, again toward a lower-energy state — in this case regularly polycrystalline (e.g., equiaxed) and defect free.

The equilibrium phases considered in Chapter 3 are of course achieved by the prolonged heating of previously metastable structures of the type considered in Chapter 4. Although heat treatment or annealing at any temperature will permit a metastable alloy to approach more closely its state of thermodynamic equilibrium, the term aging is generally understood to imply a heat treatment in the low- to moderate-temperature range. Of particular interest in this context is the aging of the martensitic and $\omega + \beta$ phases, and of course the aging of technical $\alpha + \beta$ and β alloys. It is inevitable that many of these topics have already been touched on in the previous chapters. The present chapter, therefore, restricts itself to summarizing and unifying some of the earlier discussions. Moreover, the existence of three important recent papers by WILLIAMS, on precipitation and phase transformations in Ti alloys, viz: "Kinetics and Phase Transformations (in Titanium Alloys: A Critical Review)" [WIL 73], "Phase Transformations in Ti Alloys — A Review of Recent Developments" [WIL 82[b]],*

---

*This paper, although published in 1982, was actually written in about 1975/76.

and "Precipitation in Titanium-Base Alloys" [WIL76], preempts for a time further detailed discussion of the subject.

The aging of α-phase alloys, and in particular the long-range-ordering of the DO_{19} α_2-phase Ti_3SM-base structure,* is not treated here. What will be considered though, with reference to binary Ti-TM alloys, are: (i) the transformation under aging conditions of the α' and α'' martensitic phases; (ii) aging in the $\omega+\beta$-phase field and precipitation of the isothermal ω phase; (iii) precipitation out of the β-phase field, adjacent to the $\omega+\beta$ region, of a solute-lean bcc phase designated β'—the so-called "phase-separation" reaction; (iv) the decomposition of β into $\alpha+\beta$ and practical methods of distinguishing between α and aged-ω precipitates; (v) the effects of various ternary additions on the kinetics and products of the aging reaction; and finally, (vi) consideration is given to a different type of β decomposition, this time into a pair of *equilibrium* bcc phases designated β' (solute lean, again) and β'' (solute rich), characteristic of the phase diagram of, for example, Zr-Nb. The purpose of considering this reaction is primarily to draw attention to the distinction between the $\beta \rightarrow \beta'+\beta$ nonequilibrium reaction and the reaction $\beta \rightarrow \beta'+\beta''$, where $\beta/(\beta'+\beta'')$ represents an equilibrium solid-state miscibility gap.

7.1 The Tempering of the Quenched Martensites

The term "tempering", often used to describe the aging of quenched martensites, draws by implication an analogy with the heat treatment of quenched steels to whose microstructures the term "martensite" was originally applied. In discussing the aging of Ti martensites, WILLIAMS [WIL76] has catalogued the several processes which have been identified and has drawn attention to the disagreement which has arisen over the manner in which the early stages of α'' decomposition takes place. In an interesting case study of the decomposition of hexagonal martensite, TRENOGINA and LERINMAN [TRE82] have investigated the tempering, at temperatures of from 500 to 600°C, of the $\alpha+\beta$ alloy Ti-6.5Al-3.5Mo-2Zr-0.2Si after β quenching from 1050 and 1100°C. Attempts to provide a unified mechanistic model of the decomposition processes in both α' and α'' Ti-TM alloys have been made by FLOWER, DAVIS, and WEST [FLO82][DAV79a]. Although yet to be confirmed by others, and to be applied to Ti-TM alloys in general, their approach provides a very convincing description of the manner in which the α' and α'' variants of Ti-V, Ti-Nb, and Ti-Mo decompose, depending on the concentration ranges concerned. As pointed

---

*SM infers a simple metal such as Al, Sn, etc., or mixtures of them.

out in Sects. 4.2.5(b) and (c), which dealt with Ti-Mo and Ti-Nb alloys, respectively, and illustrated by *Fig. 4-10*, the α' variant of the β-isomorphous martensites transforms directly to $\alpha + \beta$ by the nucleation-and-growth of β-phase precipitates. The reaction is fast because the α' contains a high density of heterogeneous nucleation sites and since the aging temperature, $\gtrsim M_s^{\alpha'}$, is necessarily high. Within the intermediate-concentration range for α^m transformation, a spinodal decomposition of the α'' to $\alpha''_{lean} + \alpha''_{rich}$ is supposed to take place during the quench. This process, which via the accompanying compositional modulation gives rise to the reaction $\alpha''_{lean} + \alpha''_{rich} \rightarrow \alpha + \beta$, forms the basis for the $\alpha + \beta$ cellular reaction which has been observed to take place during the aging of, for example, Ti-Mo alloys within a specified concentration range. At higher concentrations, spinodal decomposition of the α'' does not take place, and it goes directly to β during the initial stage of aging. Since the aging temperature, if near $M_s^{\alpha''}$, is not necessarily low, the product of continued aging will frequently be $\alpha + \beta$.

7.2 The Isothermal ω Phase: The Aging of ω + β-Phase Ti-TM Alloys

7.2.1 Kinetics and Morphology

Athermal ω phase has been shown to occur as a crystalline precipitate within a narrow composition range in quenched Ti-TM alloys. The sensitivity of this process to composition and experimental conditions is exemplified by the numerous studies which have been carried out on the Ti-Nb system.

While BAGARIATSKII *et al.* [BAG59], on the basis of x-ray and hardness data, claimed a composition of 18 at.% for athermal ω, a value which agrees with the results of an electron diffraction study by BALCERZAK and SASS [BAL71, BAL72], HATT and RIVLIN [HAT68] found α'' as the water-quenched product in Ti-Nb(20.7 at.%), as did HICKMAN [HIC69a] in gas-quenched ribbons of Ti-Nb(22 and 25 at.%). On the other hand, in a pair of papers discussing the structure and superconducting properties of Ti-Nb(22 at.%), BRAMMER and RHODES [BRA67] and KRAMER and RHODES [KRA67], respectively, showed that the compositional limit for athermal ω phase was already exceeded, and that the diffractographically defined "diffuse ω" with its unresolvable real-space counterpart was the quenched product. During the moderate temperature aging ($\leq 450°C$) of an alloy exhibiting the diffuse ω reflections, decomposition into the isothermal ω (solute lean) and β phases takes place. Numerous descriptions of this process accompanied by photomicrographs of the isothermal ω precipitate are to be found in the literature.

An early model [Cou69] of the physics and kinetics of isothermal ω-phase development pictured an initial *structural* transformation of the lattice into ω and β (as for athermal ω phase in *its* composition regime), followed by an exchange of solute and solvent across the ω/β interface. DE FONTAINE, PATON, and WILLIAMS [DEF71], on the other hand, suggested that the first step was a *compositional* fluctuation, followed by a structural β → ω transformation within a solute-lean zone triggered by $\frac{2}{3}\langle 111 \rangle$ longitudinal lattice vibrations, already an athermal property of the low-concentration bcc lattice (see Sect. 4.3.2). If such lattice vibrations are indeed responsible for the isothermal process, it is not surprising that during the aging of Ti-V [Mcc71] and Ti-Nb [BAL71, BAL72] the cubic or ellipsoidal, respectively, precipitates which form appear to result from the growth of clusters of $\langle 111 \rangle$ rows of particles.

Provided that the temperature is below about 400°C, after prolonged aging a metastable ω + β state is attained, characterized at a given temperature by a fixed volume-fraction and composition of the ω and β end-points [HIC69ᵃ]. A suggested metastable equilibrium phase diagram for Ti-Mo from the work of DE FONTAINE, PATON, and WILLIAMS [DEF71] is presented for further discussion in *Fig. 7-1*. Using x-ray techniques of the kind described by HICKMAN [HIC69, HIC69ᵃ], the volume-fraction of the ω phase may be obtained from the corrected relative intensities of the ω and β reflections; if the rate-of-change of lattice parameter with composition is not too small, compositions can be calculated to ±0.1 at.% from calibrated lattice-parameter measurements.

It has been determined that the *aged* product bears the same crystallographic relationship to the parent lattice as does the *athermal* ω phase [WIL76]. After sufficiently long aging times at 450 and 500°C, α-phase precipitation can be expected, as indicated in *Table 7-1*. The relative sluggishness of the decomposition process in Ti-Mo compared to that in Ti-Fe

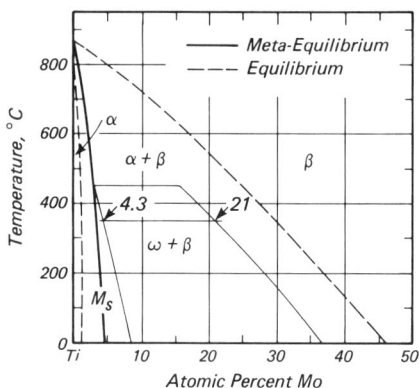

Fig. 7-1. Meta-equilibrium phase diagram for Ti-Mo indicating the ω/(ω+β) and (α+β)/β phase boundaries (fine full lines) and an M_s transus (heavy full line). The 350°C isotherm is shown intersecting the transi at 4.3 and 21 at.% Mo. Also indicated (dashed lines) is a standard equilibrium phase diagram [DEF71].

Table 7-1. Time Needed for the Appearance of α-Phase Precipitation During the Aging of Quenched Ti-TM Alloys

Alloy	Aging time (h) at temperature 400°C	450°C	500°C	Reference
Ti-V(15, 19 at.%)	20-30	<20		[HIC68]
Ti-V(19 at.%)			<4	[VET68]
Ti-V(25 at.%)	20-30	(No ω)		[HIC68]
Ti-Cr(9.3 at.%)	50			[HIC69ᵃ]
Ti-Mn(6.7 at.%)	68			[HIC69ᵃ]
Ti-Fe(6.0 at.%)	150	12	(No ω)	[HIC69ᵃ]
Ti-Nb(22 at.%)	72			[HAT68]
Ti-Mo(8 at.%)		320	50	[HIC69ᵃ]
Ti-Mo(10 at.%)		150	(No ω)	[HIC69ᵃ]

is a reflection of the great difference in the diffusion coefficients of these two solute atoms (see *Fig. 2-18*).

The compositions of the meta-equilibrium isothermal ω phases have already been given in *Table 4-2* within the context of a discussion of the limits of β-phase stability. With the aid of corresponding data for the β phase [HIC68, HIC69ᵃ], such as that presented in *Table 7-2*, a semi-quantitative metastable equilibrium phase diagram for ω + β may be assembled. *Fig. 7-2* is such a diagram for Ti-Nb. In it we note that: (i) martensitic transformation supersedes athermal ω transformation in quenched alloys (cf., for example, the results of BAGARIATSKII *et al.* [BAG59] and HATT and RIVLIN [HAT68] mentioned earlier), and (ii) at 450°C the maximal Nb concentration for isothermal ω-phase precipitation is 30 at.%. This is probably also true for lower aging temperatures, if the

Table 7-2. Nb Contents of the ω and β Phases in Aged Metastable-Equilibrium Ti-Nb Alloys [HIC69ᵃ]

Temperature of Aging: 450°C

Average Nb concentration, at.%	Aging time, h	Volume fraction of ω phase	Nb concentration, at.% ω phase	β phase
22	10	0.36	6-11	29 ± 1
	30	0.34	5-10	30 ± 1
	50	0.33	6-11	31 ± 1
25	10	0.25	7.5-12	30 ± 1
	24	0.26	7.5-12	30 ± 1

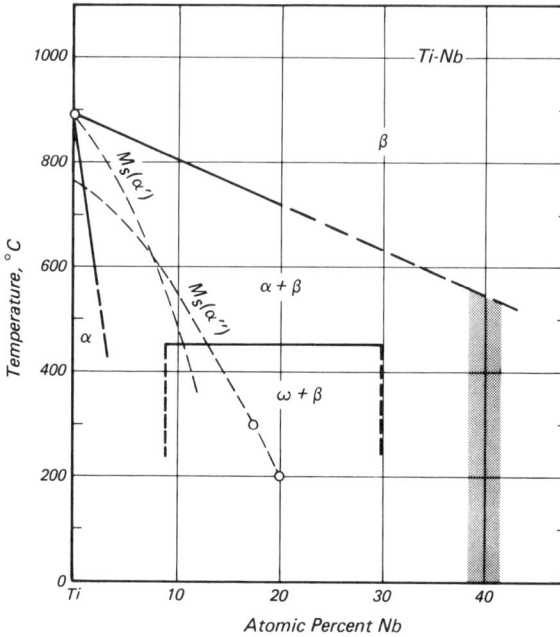

Fig. 7-2. The locations of the α- and β-equilibrium transi, the $M_s(\alpha')$ and $M_s(\alpha'')$ transi, and the regimes of occurrence of athermal and isothermal ω phases in Ti-Nb. *Data sources:* α and β transi, [MOL65, p. 20], see also *Fig. 3-10; M_s* transi, [JEP70] and [FLO82]; $\omega + \beta$ phase data, [HIC69[a]].

results of the isothermal aging studies of Ti-V(15 at.%) [HIC68] and Ti-Cr(9.3 at.%) [HIC69[a]], at temperatures between 300 and 400°C (which yielded practically vertical $(\omega + \beta)/\beta$ transi), can be accepted as having general significance. The isothermal precipitate particles assume one of two types of morphology — cubic or ellipsoidal — depending on the linear lattice misfit $(V_\omega - V_\beta)/3V_\beta$, where V represents the unit-cell volume divided by the number of atoms per unit cell [HIC69[a]]. If this is large (1-3%), as in Ti-V, Ti-Cr, Ti-Mn, and Ti-Fe [HIC69], minimization of elastic strains in the cubic matrix dictates a cubic morphology, *Fig. 7-3*. If the misfit is small (<0.5%), as in Ti-Mo and Ti-Nb [HIC69, HIC69[a]], the morphology is dominated by surface-energy considerations, leading to the ellipsoidal particle shape depicted in *Fig. 7-4*. The influence of misfit on ω-particle morphology has been graphically demonstrated by WILLIAMS *et al.* [WIL71] in experiments in which the addition of 5.5 at.% Zr to Ti-V(20 at.%) resulted in a decrease in the misfit from 1.5-2.0% to ~0.25% and caused the precipitate shape to change from cubic to ellipsoidal.

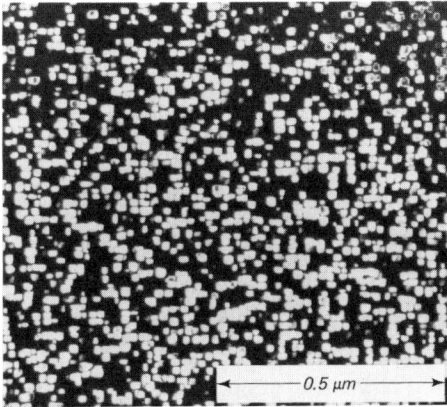

Fig. 7-3. An example of cubic ω phase. *Specimen*: Ti-10Fe [WIL 73 *(corrected)*]. Micrograph courtesy of J.C. Williams (Carnegie-Mellon University); copyright © 1973, Plenum Publishing Corporation, reprinted with permission.

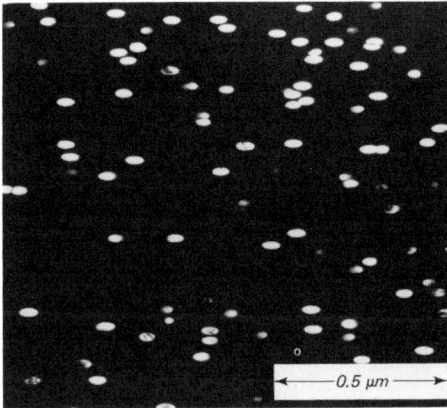

Fig. 7-4. An example of ellipsoidal ω phase. *Specimen*: Ti-11.5Mo-6Zr-4.5Sn [WIL 73 *(corrected)*, WIL 76]. Micrograph courtesy of J.C. Williams (Carnegie-Mellon University); copyright © 1973, Plenum Publishing Corporation, reprinted with permission.

7.2.2 Influence of Aging on Alloy Properties

Phenomenological studies of the effects of aging on the properties of $\omega + \beta$ Ti-TM alloys were conducted almost thirty years ago by FROST *et al.* [FRO 54] and BROTZEN *et al.* [BRO 55]. Their results, along with numerous others, have already been thoroughly reviewed by McQUILLAN [McQ 63, pp. 51-7] and MARGOLIN and NIELSEN [MAR 60, pp. 257-62].

It was the hardening and embrittling properties of ω phase that originally led to its discovery [PAR 53]; subsequently it was found that the hardness, already characteristic of quenched $\omega + \beta$-phase samples, increased with aging time [FRO 54][BRO 55]. During aging, the tensile and yield strengths increased and the ductility (elongation at fracture) decreased.

Upon overaging, during which the ω phase dissolves and is replaced some-how by α phase, the ductility is restored. Most of the common transition elements (Nb is an exception) decrease the lattice parameter of the bcc phase; thus, during the isothermal aging of $\omega + \beta$, the lattice generally shrinks as the β component becomes enriched in solute. If the aging temperature is increased and the alloy overaged to $\alpha + \beta$, the lattice may expand to accompany a re-adjustment of the volume fraction of β phase to its equilibrium value [BRO55][MAR60]. More recently, using Ti-Cr as a candidate system, POLONIS and co-workers have conducted an exhaustive investigation of isothermal aging and its effects on microstructure [CHA73^a, CHA78], electrical conductivity [CHA74], and superconducting properties [LUH69, LUH70^a]. During the same period of time, COLLINGS and Ho applied physical-property measurement techniques to the study of isothermal $\omega + \beta$ aging in several previously β-quenched binary Ti-TM alloys. Techniques employed were: magnetic susceptibility (Ti-Fe) [Ho73], (Ti-V) [COL75^b]; electrical resistivity (Ti-Mo) [COL72], (Ti-V) [COL74]; low-temperature specific heat (Ti-Fe) [Ho73], (Ti-V) [COL75^c, COL82^b], (Ti-Mo) [COL72^a, COL82^b][Ho73^a].

Of particular interest are the results of all the electrical resistivity investigations. The temperature dependences of electrical resistivity, $d\rho/dT$, of as-quenched and quenched-plus-aged binary Ti-TM alloys have been studied by both of the above-mentioned research groups, and by a dozen or so previous workers whose results were briefly reviewed by CHANDRASEK-ARAN et al. [CHA74]. For alloys both within and outside the composition range for the occurrence of athermal ω phase, $d\rho/dT$ is negative. Electron scattering from thermally reversible athermal ω phase would obviously provide a mechanism for this effect. However, the observation of negative $d\rho/dT$ in quenched Ti-Cr(20 at.%), which on cooling in the electron microscope to $-180°C$ revealed no ω-phase reflections, suggested that the anomalous resistivity temperature dependence was associated with the instability of the β lattice itself. As shown in *Fig. 7-5*, for a series of quenched-and-aged Ti-Cr(10, 13, 15, 20 at.%) alloys during aging at 300°C for more than 16 h, $d\rho/dT$ sigmoidally approached a common positive (normal) value as the precipitation of isothermal ω phase took place.

7.3 β-Phase Separation

7.3.1 Occurrence of the β' Precipitate

When the temperature is too high [LUH70][WIL71] or the alloys too con-centrated [WIL76] to support ω-phase precipitation, a solute-lean bcc pre-cipitate, designated β', separates out. The relationship in temperature/composition space between the metastable $\omega + \beta$- and $\beta' + \beta$-phase fields is

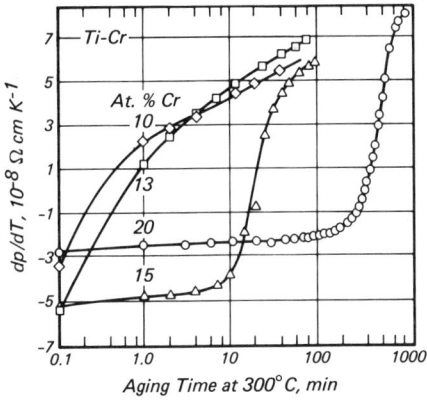

Fig. 7-5. Influence of aging time at 300°C on the resistivity-temperature-dependence of initially quenched Ti-Cr alloys [CHA74].

indicated schematically in *Fig. 7-6*, which itself is based on *Fig. 4-1* and *Fig. 7-2* and some suggestions by WILLIAMS *et al.* [WIL71]. As with $\omega + \beta$, the $\beta' + \beta$ mixed phase is metastable; but unlike ω, which can be generated by a displacement wave in a virtual crystal if need be, the β' precipitate stems from the chemical differences between the solute and solvent atoms. Thus, the $\beta \to \beta' + \beta$ "phase-separation" reaction, as it is called, is a clustering reaction characteristic of alloy systems which show positive heats of mixing [CHA72a] or equivalent manifestations of a tendency for the alloying constituents to un-mix. Partly as a consequence of the time-temperature conditions for β' formation and ω-phase precipitation, the former (or at least the precursor stages of it) was for a time confused with ω phase. This is no longer the case, the status of β' as a metastable bcc precipitate now being well established [WIL73, WIL76].

The $\beta \to \beta' + \beta$ reaction has been studied in considerable detail using Ti-Cr [LUH70][NAR71] and Ti-Mo [KOU70, KOU70a][NAR70][CHA72a] as model systems, although its appearance in more complicated alloys (e.g., Ti_{72}-Zr_8-V_{20}) is well documented [WIL73, WIL76]. Typical examples of β'

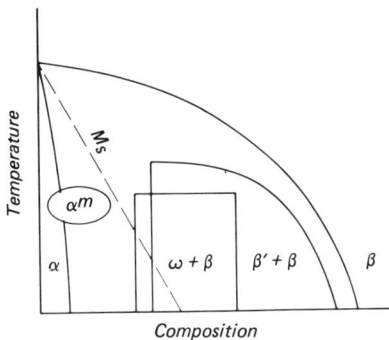

Fig. 7-6. Schematic representation of the locations of the two metastable phases $\omega + \beta$ and $\beta' + \beta$ within the equilibrium $\alpha + \beta$-phase field in a typical Ti-TM alloy [WIL71].

precipitate morphology are given in *Fig. 7-7*. It is interesting to note that a structure very much similar to that illustrated has been discovered in the Soviet alloy 65 BT (composition about Ti_{40}-Zr_7-Nb_{53}) after water quenching from 1250°C and aging 2h/700°C [BUY70]. Evidence for phase separation in Ti-35Nb (22 at.%) has been offered by MENDIRATTA *et al.* [MEN71]. Ti-15Cr, when quenched, appeared to be single-phase bcc [NAR71] and after $5\frac{1}{2}$min/300°C exhibited the characteristics of diffuse ω phase. But after holding the temperature at 450°C for 30 min, β' precipitation in the form of thin discs (~0.1 μm across) lying parallel to the {100} planes of the matrix [LUH70] were clearly discernible when photographed at a magnification of 52,000×. In this case apparently, for aging times longer than 90 min at 450°C, α precipitation commenced at the β'/β interfaces [LUH70].

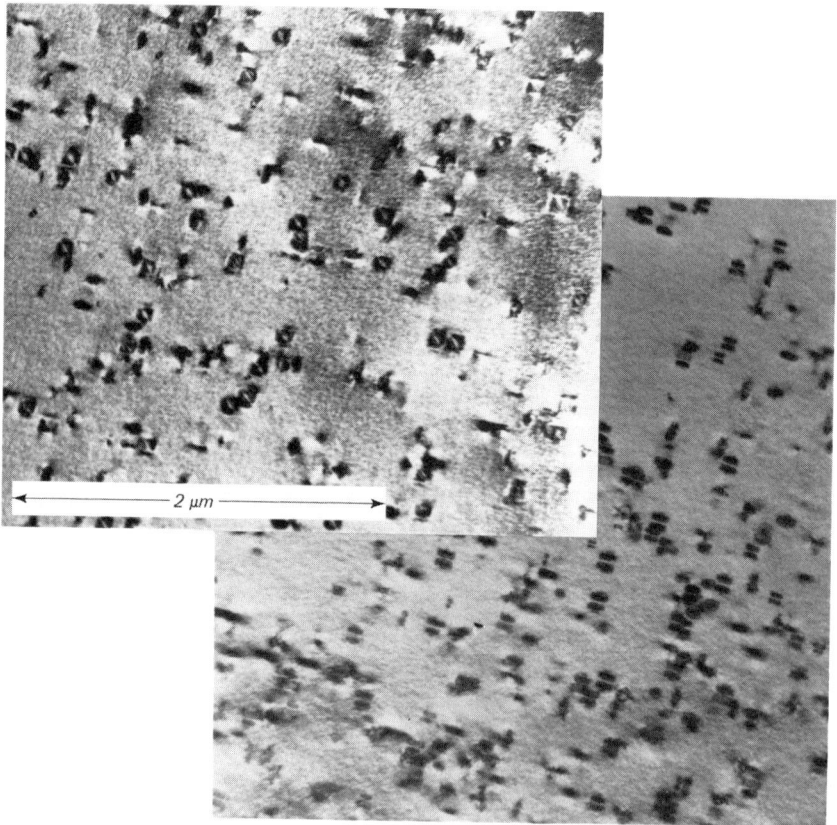

Fig. 7-7. An example of β' precipitation in a β matrix; two orientations of the same sample are represented. *Specimen*: Ti_{72}-V_{20}-Zr_8 [WIL73]. Micrographs courtesy of J.C. Williams (Carnegie-Mellon University).

Using Ti-V and Ti-Mo alloys as host systems, WILLIAMS and colleagues [WIL71] have considered the influences of temperature, composition, and third-element addition on the relative stabilities of the ω and β' phases and the precipitation processes that are associated with them. An interesting conclusion that can be drawn from their results is that α stabilizers such as Al, Sn, and O, as well as Zr, when added to those bases in sufficient quantity, increased the stability of the bcc lattice, thereby suppressing ω-phase formation in favor of β-phase separation. It was suggested on those grounds that the presence of Al in the β-stabilized commercial alloy Ti-8Mo-8V-2Fe-3Al and Al plus Zr in Ti-3Al-8V-6Cr-4Mo-4Zr (β-C) would stimulate the phase-separation reaction as an intermediate step toward a very uniform distribution of α phase under suitable heat treatment (see below). Although their results were unable to be substantiated by later measurements [MOR80], RHODES and PATON [RHO77] amply confirmed the prediction of WILLIAMS et al. [WIL71] by detecting β' precipitation in β-C in response to aging at temperatures of 350°C (precipitation after 24 h) to 500°C (precipitation after $\frac{1}{2}$ h).

7.3.2 Thermodynamics of the Phase-Separation Reaction

Although early studies indicated that Ti-Mo was a short-range-ordering system, the converse is now known to be true [CHA72ᵃ]. HOCH and VISWANATHAN [HOC71], using a relative vapor pressure technique, have determined the activity of Mo in Ti and shown it to be positive; and using the more direct technique of low-angle x-ray scattering, GEHLEN has shown that Ti-Mo is a clustering system [HO69]. Consequently, in a temperature-composition zone lying just outside the $\omega + \beta$-phase field, the tendency for compositional modulation (which underlies the establishment of the equilibrium $\alpha + \beta$ field itself) stimulates in Ti-20Mo, for example, at 450°C [CHA72ᵃ] a $\beta \rightarrow \beta' + \beta$ phase-separation reaction similar to that just noted in Ti-14Cr. These examples serve to illustrate the general clustering tendencies of solid solutions of Ti with transition metals; tendencies which may be gauged, for example, by the algebraic sign of the thermodynamic interaction parameter Ω_{ij}. A set of interaction-parameter values for some representative transition metals and simple metals against Ti has been assembled by COLLINGS and GEGEL [COL75ᵃ]. As pointed out by those authors, the simple metals, which yield negative interaction parameters, give rise to α-stabilized, short-range-ordered systems when alloyed with Ti. Conversely the transition elements, which have positive interaction parameters, are clustering-type β stabilizers. Although KOUL and BREEDIS [KOU70, KOU70ᵃ], NARAYANAN and ARCHBOLD [NAR70], and GULLBERG et al. [GUL71] used a bimodal free-energy/composition curve to treat phase separation as an equilibrium event, it must be remembered that $\beta' + \beta$ is a

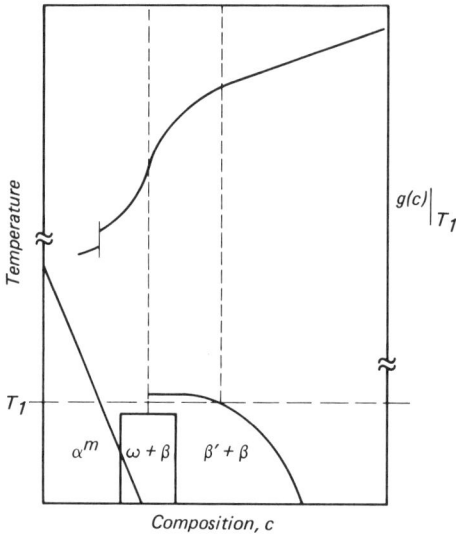

Fig. 7-8. Schematic metastable equilibrium phase diagram for quenched-and-aged Ti-TM alloys accompanied by a postulated free energy, g, versus composition, c, curve for some temperature, T_1, showing a segment with negative curvature corresponding to the phase-separation region of the phase diagram.

metastable mixture with regard to which only an inflected free-energy/composition relationship (implying spinodal decomposition wherever $\partial^2 g/\partial c^2 < 0$) is a sufficient description, *Fig. 7-8*. A bimodal free-energy/composition curve, to be considered below, implies the existence of a miscibility gap in the β field. Whereas this is not a feature of any of the Ti-TM phase diagrams, the Zr-Nb system does possess an *elevated temperature* field representing, in equilibrium, two immiscible β phases designated β' (solute lean, as before) and β'' (Nb rich). The equilibrium $\beta' + \beta''$ phase is to be discussed below in connection with the Ti-Zr-Nb system.

7.4 α-Phase Precipitation from β Alloys

7.4.1 Direct Precipitation

Outside the composition and temperature ranges of the ω or β' phases, represented schematically in *Fig. 7-6*, but still within the equilibrium $\alpha + \beta$ field, α-phase precipitation must take place "directly" from the β phase. Mechanisms of direct α-phase precipitation have been discussed in considerable detail by WILLIAMS [WIL73, WIL76]. A well-known morphology is the Widmanstätten α phase which, since it forms as platelets parallel to $\{110\}_\beta$, *Fig. 4-6*, assumes under optical magnifications a characteristic basket-weave-like structure. The interface between the α plates and the matrix is then the source of further, apparently athermal [WIL76], α

phase. Other forms of direct α-phase precipitation which have been discussed by WILLIAMS take on the form of large clumps >1 μm across [WIL73], or dispersions of fine particles ~400 Å in diameter [WIL76]. Also of great interest when considering the morphology and kinetics of α-phase precipitation are the ways in which α phase nucleates heterogeneously from "defects" such as grain boundaries, dislocations, β'- and ω-phase precipitates.

7.4.2 Precipitation from the β' + β Phase

In Ti-15Cr, according to LUHMAN [LUH70], prolonged aging in the β' + β field (e.g., >90min/450°C) resulted in a nucleation of α phase at the β'/β interfaces. On the other hand, WILLIAMS et al. [WIL71] were careful to point out that for the Ti-Mo-Al alloy which they studied nucleation of the α phase took place within the β' region, indicating that in this case the composition difference between precipitate and matrix was the controlling factor. With only two examples, insufficient information is available to establish any systematics concerning the relative roles of lattice mismatch and composition gradient in the β'→α nucleation reaction. An ability to nucleate α seems to be a general property of β' precipitates, but the subsequent growth of the α phase may be either acicular (a fine dispersion of star-like objects) or globular [WIL76].

7.4.3 Precipitation from the ω + β Phase

As indicated in *Table 7-1*, the over-aging of ω + β generally results in α-phase precipitation. Several precipitation modes have been identified, none of which involves a direct conversion of ω phase to α. As a matter of fact, in only one alloy system, Ti-V, is the ω-phase precipitate itself involved in the conversion process; in it BLACKBURN and WILLIAMS [BLA68] have demonstrated the growth of α plates at the ω/β interfaces. The exceptional role played by Ti-V in this regard is not surprising in view of the fact that the ω phase in Ti-V possesses one of the largest lattice misfits [HIC69]. In Ti-Mo and Ti-Nb, which support the lowest misfit ω phases, α precipitation takes place elsewhere than near the ω-precipitate site. In Ti-Mo [BLA68], large-scale lamellae of α and β phases have been observed growing from a grain boundary, and α plates have been noted to nucleate at dislocations. Similarly, in Ti-Nb [BRA67] α phase nucleating at β grain boundaries grew to consume the β and ω phases.

With alloys such as Ti-Nb, lattice defects not only act as nucleation sites for α precipitation, they also provide strain energy to accelerate the transformation process. The influence of prior deformation on the aging process is to be considered below.

7.5 Down-Quenching and Up-Quenching: ω Reversion

By taking an alloy, previously aged isothermally in the $\omega + \beta$ field and, by raising the temperature, placing it in the $\beta' + \beta$ field, it is possible to exert some control over the β-phase separation reaction [LUH70]. The isothermal ω phase "reverts" at the higher temperature to a β phase leaner in solute content than the matrix. The reaction is not reversible, the β-phase precipitates remaining in place as the alloy is returned to room temperature. The new phase could be thought of as β', since it is stabilized by the inherent tendency of phase separation (clustering) to occur in the parent β alloy [LUH70]. In studies of Ti-15Cr, LUHMAN [LUH70] observed that a larger number-density of finer precipitates resulted from up-quenching to 450°C for $2\frac{1}{2}$ min from isothermal $\omega + \beta$, aged at 300°C, than were found in the same alloy down-quenched to 450°C from the β field. Since as before, β' provides a site for the nucleation of α precipitates, these too can be produced as a fine dispersion as a result of this duplex aging treatment which, therefore, affords a means of obtaining the strength which is characteristic of fine uniform precipitation, but which is unencumbered by the embrittling effect of ω phase [LUH70]. It is useful to note in conclusion that, even prior to α-phase precipitation, the modulated $\beta' + \beta$ structure itself is associated with enhanced strength compared to that of the quenched homogeneous β alloy, and greater ductility than that of $\omega + \beta$ [POL71].

7.6 Effect of Third-Element Additions on Precipitation in Quenched and Aged Titanium-Transition-Metal Alloys

Current fundamental understanding of the manner in which third-element additions influence the $\omega + \beta$ and $\beta' + \beta$ decomposition reactions stems from an investigation by WILLIAMS, HICKMAN, and LESLIE [WIL71] into the effects of Zr, Al, Sn, and O additions on precipitation in Ti-V(20 at.%) and Ti-Mo(6 at.%). Their results can be readily appreciated in the light of the preceding sections. As before, precipitation effects in the $\omega + \beta$ and $\beta' + \beta$ regimes, respectively (see *Fig. 7-6*), will be treated separately.

7.6.1 The Ternary ω + β-Phase Regime

Ternary additions of Al and O reduced the time of stability of the ω phase during aging by promoting early precipitation of α phase. In Ti-V(20 at.%) for example, the addition of 3 at.% Al reduced the time required at 400°C to produce a given volume-fraction of α precipitation from 100 to

24 h; and whereas in the binary alloy, grain-boundary nucleation was preferred, Al in solid solution promoted the formation of a uniform dispersion of precipitates. $Ti_{74.5}$-V_{20}-$Zr_{5.5}$, although within the $\omega + \beta$ field, is close to its boundary with $\beta' + \beta$ (to be discussed below). Thus, as compared to the results of aging in the binary system, that alloy yields a low volume-fraction of isothermal ω phase. Furthermore, as a consequence of the now relatively small lattice misfit, the precipitate is ellipsoidal rather than cubic.

7.6.2 The Ternary β' + β-Phase Regime

As was noted above in connection with binary alloy decomposition, aging in the $\beta' + \beta$ regime, which lies outside the $\omega + \beta$ field (*Fig. 7-6*), eventually results in β'-nucleated α precipitation. Additions to binary Ti-TM alloys of solutes such as Zr and Sn tend to stabilize the bcc structure when dissolved in it.* This is not to say they are "β stabilizers" in the conventional sense; but, rather, stabilizers of the bcc lattice against the ω instability. Zr and Sn thus impose constraints on the $\omega + \beta$ regime to such an extent that alloys such as Ti_{72}-V_{20}-Zr_8 and Ti_{84}-Mo_{10}-Sn_6 are left outside it and consequently lie in the $\beta' + \beta$ field. Aging then brings about the phase-separation reaction which results in the precipitation of β', the site for eventual precipitation of finely dispersed α phase.

Finally, it is important to recall that irrespective of whether the alloys were in the $\omega + \beta$ or $\beta' + \beta$ fields, the electron diffraction patterns were characterized by the reciprocal-lattice "streaking" effect discussed earlier in connection with the $\langle 111 \rangle_\beta$ longitudinal displacement-driven ω-phase reaction. These results confirm that whether or not an ω or a β' transformation is eventually to take place, bcc transition metal alloys (along with other classes of bcc metals; see for example SMITH *et al.* [SMI76]), support to some degree the $\frac{2}{3}\langle 111 \rangle$ soft phonon.

7.7 β-Phase Immiscibility

In contrast to the $\beta \rightarrow \beta' + \beta$ phase-separation reaction discussed above is a second type of β-decomposition process with which, in one way or another, it has been frequently confused. Below a transus delineating the upper boundary of a region referred to as a "miscibility gap" a previously

---

*The solution strengthening of both α-stabilized and β-stabilized Ti-TM alloys has been discussed by COLLINGS and GEGEL [COL75a][GEG73a].

homogeneous single-phase-β solid solution decomposes into a *thermodynamically stable* pair of bcc phases — a solute-lean β' and a solute-rich β''.

An ideal example of $\beta \rightarrow \beta' + \beta''$ decomposition is to be found in the system Zr-Nb, whose equilibrium phase diagram is comparable to that of the general Ti-TM eutectoidal alloy system (*Fig. 3-2(b)*) with the Ti replaced by Zr, and the intermetallic compound (TiTM$_x$, or simply γ) replaced by β''. Indeed, common to precipitation from both the aged $\beta' + \beta''$ (and in fact $\alpha + \beta''$) and the aged $\alpha + \gamma$-phase fields is the lamellar, cellular, or pearlitic morphology characteristic of a classical discontinuous growth (see [CHR65, p. 472]). Precipitation from within the $\beta' + \beta''$ field does, however, possess an additional feature (to be discussed in detail below) not shared by the $\alpha + \gamma$ precipitation process.

Other examples of β-phase immiscibility are to be found in ternary alloys based on Zr-Nb, such as the Ti-Zr-Nb system of *Fig. 3-22* and *Fig. 7-9*. The development of the equilibrium $\beta' + \beta''$ region from the Zr-Nb edge of the composition triangle is exemplified in *Fig. 7-9*, where it is juxtaposed against the appropriate isothermal section of the Zr-Nb

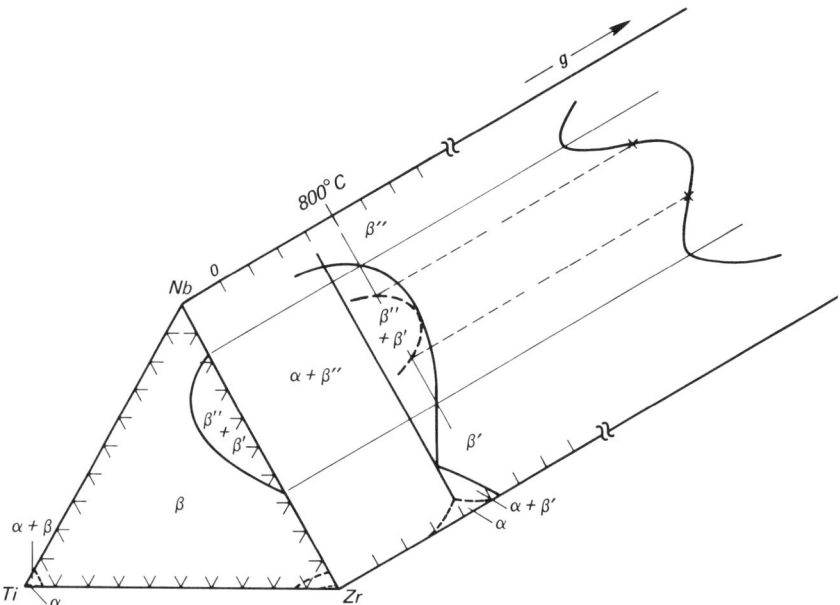

Fig. 7-9. Binary equilibrium phase diagram for Zr-Nb [Lov66] projected from a side of the ternary (at.% linear) Ti-Zr-Nb 800°C-equilibrium phase diagram [Doi66]. The two minima in the bimodal $g(c)$ curve define a region of equilibrium β-phase immiscibility, $\beta' + \beta''$, within which is a region of spinodal decomposition delimited by the points of inflection on $g(c)$.

binary phase diagram. Associated with the binary diagram is a bimodal free-energy curve representing the existence *in equilibrium* of two β solid solutions. Such a curve has been incorrectly applied to the $\beta' + \beta$ *metastable* situation [KOU70, KOU70^a][NAR70][GUL71]. Thermodynamic equilibrium requires a free-energy versus composition minimum, as distinct from the inflected type of free-energy versus composition relationship which characterizes the metastable phase-separated condition of *Fig. 7-8*. But in contrast to $\alpha + \beta$ phase equilibrium which is described by two independent free-energy parabolas, equilibrium phase-separation with its absence of structural change must be described by a *continuous* curve with two minima. An important corollary to this is the necessary presence of an intervening maximum with its associated description in free-energy terms: $\partial^2 g/\partial c^2 < 0$. In other words, away from the edges of the two-phase field, the $\beta \to \beta' + \beta''$ equilibrium process is preceded by spinodal decomposition. In *Fig. 7-9* two points on the spinodal (indicated by the dashed curve on the phase diagram) are determined by the two points marked "×" on the free-energy (g) versus composition (c) curve for which $\partial^2 g/\partial c^2 = 0$. Precipitation phenomena interpretable in this manner have been seen during metallographic studies of quenched-and-aged Ti_{10}-Zr_{40}-Nb_{50}. During the early stages of its aging, according to KITADA and DOI [KIT70], a very fine precipitate developed, presumably spinodally, throughout the grain interior such that after 10h/700°C the hardness had increased from 280 to 338 kg mm^{-2}. In the meantime, a lamellar or pearlitic precipitate of alternate β' and β'' regions, of much lower hardness ($\sim$200 kg mm^{-2}), had begun to develop outward from the grain boundaries.

Part 2: The Kinetics of Precipitation

The equilibrium and metastable phases which appear during the aging of quenched β-Ti-TM alloys were considered in Chapters 3 and 4 and in Part 1 of this chapter. These *same phases* appear during the aging of deformed alloys but at *different rates* determined by the degree of strain and the level of retained stress. The following sections, therefore, deal with some of the factors which control the *kinetics* of the aging process.

7.8 The T-T-T Diagram

The traditional descriptor of transformation kinetics is the time-temperature-transformation (T-T-T) diagram in which phase boundaries are plotted semilogarithmically against temperature and time. Sets of the typically C-shaped T-T-T diagrams for the systems Ti-V, Ti-Cr, Ti-Mn, Ti-Fe, Ti-Ni, and Ti-Cu are to be found in [MAY61] and the further references contained therein. Such diagrams are usually obtained by detecting iso-thermal transformations in samples which had initially been quenched from single-phase elevated-temperature fields to the temperatures of observation. It must be borne in mind, however, that such curves will not be coincident with those obtained by performing isothermal aging on specimens previously quenched to room temperature [MAY61], although they will be of the same general form. The characteristic C-shape of the T-T-T diagram, *Fig. 7-10*, demands an explanation. Microstructural studies of samples quenched from various regions of the temperature-time diagram indicate that, at high temperatures, transformation begins at lattice defects such as grain boundaries and grows into the grains (discon-tinuous precipitation), while at the low-temperature extreme the precip-itation tends to take place uniformly throughout the grains (continuous precipitation). These effects result from the fact that the driving force for nucleation (the Δg between the pairs of phases involved) scales with some measure of the degree to which the energy of the quenched alloy differs from its equilibrium value. Once nucleated, the precipitate growth rate increases with temperature; consequently, the transformation rate, which is proportional to the product of nucleation rate and growth rate, is maxi-mum at some intermediate temperature [CHR65, p. 453]. The separation of the C-curve into a low-temperature nucleation-controlled region and a high-temperature growth-controlled region leads to an understanding of the manner in which stress, strain, and interstitial atoms influence its shape. Thus, if the density of nuclei for transformation is increased, by deformation or some other means, the effect will be to increase the trans-formation speed at high temperatures, where recovery tends to eliminate nucleation sites, much more than at low temperatures, where the driving force for nucleation is high and the alloy already well supplied with nucle-ation sites. The result of combining a more nearly even distribution (with temperature) of nucleation rates with a growth rate that increases with

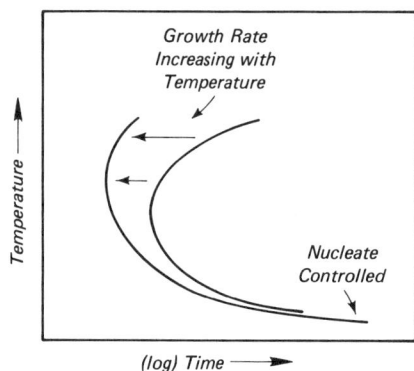

Growth Rate
Increasing with
Temperature

Temperature ⟶

Nucleate
Controlled

(log) Time ⟶

Fig. 7-10. Schematic representation of factors which control the form of the T-T-T curve. For a given nucle-ation rate the indicated increase in reaction rate (= product of nucle-ation and growth rates) has the effect of raising the "nose" of the T-T-T curve and giving it a "forward tilt".

temperature is to shift the C-curve toward the temperature axis by an amount which increases uniformly with temperature; in doing so, this also has the effect of raising the temperature of the "nose" in the manner depicted in *Fig. 7-10*. Several specific examples of this effect are given below.

The (time, temperature) data-pairs representing some selected fraction of material transformed (e.g., 5% for the start of the transformation and 95% for its end) may be obtained either at-temperature, using a diagnostic technique such as dilatometry or electrical resistometry [Hor73], or from interpretation of quenched structures using hardness or metallography, which in any case should always be employed as a semidirect observation and control. Two mechanical variables—stress level and degree of plastic deformation, both of which have strong influences on the *decomposition kinetics* of β-Ti-TM alloys—are considered below. Also to be discussed are the influences of third-element additions. These have, of course, just been considered in connection with aging transformations in *quenched* β alloys, but inasmuch as substitutional elements have a pronounced tendency to accumulate at dislocations and grain boundaries during aging, they also play an important role in the phase decomposition of plastically deformed alloys.

7.9 Influence of Stress on the Transformation

Studies by GOLDENSTEIN *et al.* [Gol59] on a series of Ti-Cr alloys, which underwent the eutectoid transformation $\alpha + \beta \rightarrow \alpha + TiCr_2$, provide a useful example of stress-assisted* decomposition. Application during isothermal aging of a relatively low level of stress (capable of producing 1% plastic strain in 1000 h) reduced the time needed to complete the eutectoid reaction in Ti-12Cr at 550°C from 200 to 32 h. In a like manner, the so-called "internal stress fields" (see TYSON [Tys75]) of lattice defects may also speed the transformation process.

7.10 Heavy Plastic Deformation and Its Properties Under Heat Treatment

It is common practice in alloy preparation laboratories to heavily deform, if possible, the as-cast billet to accelerate the equilibration reaction during a subsequent homogenization heat treatment. This section considers the effect of aging on microstructure and precipitation in heavily

---

*The term "stress" as used here implies elastic strain as distinct from the (heavy) plastic strain to be discussed later.

cold-worked β-Ti-TM alloys, particular emphasis being placed on the influence of deformation by wire drawing.

7.10.1 Influence of Aging on the Fibrous Cell Structure

During the early stages of moderate-temperature aging, the intracell dislocations migrate to the cell walls, then at higher aging temperatures extremely rapid cell growth sets in. In what follows the mechanism and kinetics of this process will be discussed.

(a) Dislocation Motion. A strongly cold-worked alloy with subband diameters in the range 100-700 Å may contain a dislocation density, within the cells, of about 10^{12} cm^{-2} [PFE68].* During moderate-temperature heat treatment ($\leqslant 300°C$ [NEA71]), migration of the dislocations to the cell walls begins to take place by some sort of thermally activated slip process. Evidence for the occurrence of this perfectly general phenomenon, which is discussed in standard works on dislocations (e.g., [FRI64]), is to be found in the sharpening of the x-ray line pattern [CHA70]. Dislocation density is still high after 1h/385°C in the large cells which result from low levels of cold deformation but much lower, according to NEAL et al. [NEA71], in the narrow cells of the heavily reduced materials with their higher levels of strain energy and shorter path lengths. Viewed in cross-section the cells possess curved boundaries and are irregular in size; then during the early stages of heat treatment the cells walls become straight [NEA71] as in the early stages of classical polygonalization [FRI64] [CHA70]. Referring to the 1-h annealing of heavily cold worked (5×10^4:1) Ti-58Nb, the intracell dislocation density is lowest at 400°C [NEA71], but the cell structure itself is on the threshold of growth; at about 380°C the deformation-cell structure seems to be moderately stable [NEA71].

(b) Cell Growth. Cell growth during the heat treatment of deformed Ti-Nb alloys has been considered briefly by several authors [REU66] [NAR66][PFE68][BAK70] and investigated quantitatively by NEAL et al. [NEA71] and ARNDT and EBELING [ARN75][WIL75], the results of whose work on this subject are combined in Figs. 7-11(a) and (b). During annealing at 385°C the cell diameter increased from 443 to 536 Å in the first hour, and then remained stable at 530 ± 6 Å at least for another 24 h [NEA71]. At 500°C the cells grew to diameters of more than 2000 Å after just a few hours [ARN74], Fig. 7-11(a). Although at temperatures less than 385°C the cells grow (<20%) to stable size within about 1 h, at 400°C and above the cell size is strongly dependent on both aging time and temperature, Fig. 7-11(b).

---

*A cell of diameter 500 Å (area = 1.96×10^{-11} cm^2) would enclose 20 dislocations.

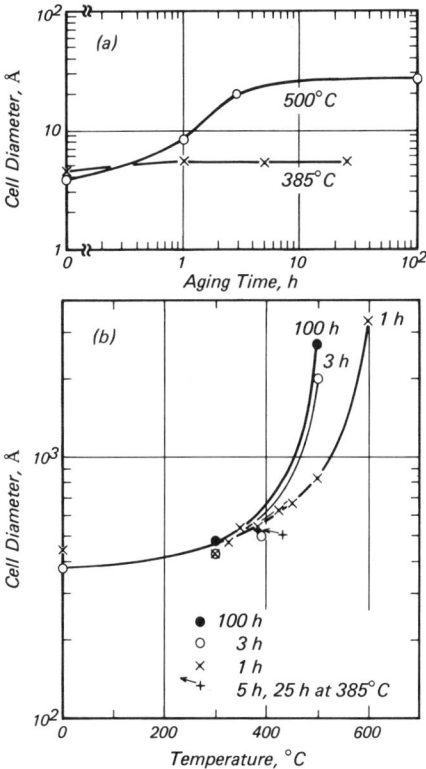

Fig. 7-11. Deformation-cell growth in β-Ti-Nb as a function of (a) time and (b) temperature. *Data sources*: ARNDT *et al.* [ARN74] (prior deformation 99.98% ($\circ$)) and NEAL *et al.* [NEA71] (prior deformation 99.998% ($\times$)).

7.10.2 Influence of Heavy Deformation on the Kinetics of Precipitation

The literature abounds with descriptions of the microstructures and precipitates which occur during the aging of cold-deformed alloys. For the purpose of this discussion, the alloy system Ti-Zr-Nb has been selected on account of its ability to support several decomposition modes.

As indicated in *Fig. 3-22*, the Zr-Nb-rich members of the system may exist in any one of four equilibrium-phase fields: (i) $\alpha + \beta$, (ii) $\alpha + \beta' + \beta''$, (iii) the miscibility gap region, $\beta' + \beta''$, or (iv) the single-phase β region at elevated temperature. *Fig. 7-12*, constructed from electrical resistivity data, describes the kinetics of transformation of previously deformed $Ti_{4.4}$-$Zr_{40.7}$-$Nb_{54.9}$ into two of these regions, viz: (a) into discontinuous $\beta' + \beta''$ from pre-deformation equilibrium in the β regime at 1200°C, and (b) into $\alpha + \beta$ from a pre-deformation equilibration in the miscibility-gap region at 700°C, as functions of percent area reduction by cold drawing. Obviously the deformation introduced by the cold work has a major

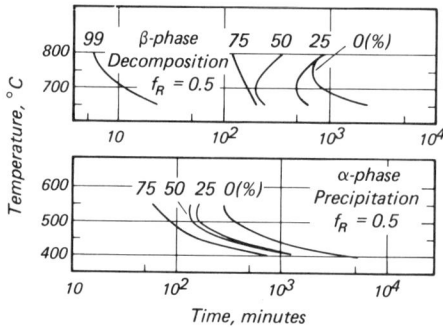

Fig. 7-12. T-T-T diagrams representing the β-phase decomposition and α-phase precipitation kinetics in $Ti_{4.4}$-$Zr_{40.7}$-$Nb_{54.9}$ cold drawn to the percentages indicated. The degree of completeness of the transformation, in resistometric terms, is $f_R = 0.5$ (where f_R is the fractional change in electrical resistance) [Soe69].

influence on the transformation speeds. The effect of deformation is greatest toward the upper temperature limits of each range of stability; thus, each C-curve is shifted, and tilted toward the temperature axis, and the "nose"-temperature is raised.

7.11 The Influences of Interstitial-Element Additions on the Kinetics of Precipitation

To be considered in this section are the ways in which the interstitial elements B, C, N, and O influence phase decomposition during the aging of cold-worked β-Ti-TM alloys. Precipitational effects associated with the presence of O in Ti-Nb(35 at.%) and Ti-Nb(46 at.%) have been discussed by Witcomb and Dew-Hughes [Wit73] and Reuter et al. [Reu66], and the influence of N on Ti-56Nb has been considered by Baker [Bak70]. The kinetics of O diffusion has been discussed by Charlesworth and Madsen [Cha70], and the influence of O on the kinetics of β-phase decomposition by DeLazaro and Rostoker [Del53] and Williams et al. [Wil71].

Interstitial elements are extremely potent strengtheners of Ti alloys as a consequence of their ability to pin dislocations (see Sect. 5.8.1). Although it is unlikely, in view of the high mobility of interstitials, that dislocation pinning takes place *at temperature*, several studies of O-doped Ti-39Nb [Com67] have indicated that coarser deformation substructures can be expected to result from the presence of strengthening centers (see also Sect. 6.4.2).

The diffusion of interstitial and substitutional elements in Ti has been thoroughly reviewed by Zwicker [Zwi74, Chap. 5], some of whose collected data have already been used in the construction of *Fig. 2-18*, which describes the relative rates of diffusion of the 3d and some 4d transition elements in β-Ti. Characteristic data (i.e., frequency factor and molar activation energies) describing the diffusion of C, N, O, and H in α- and β-Ti are available in the literature [ZWI74, p. 104]; those for O in β-Ti are reproduced in *Table 7-3*. Their reliability is obviously open to question; for example, they lead to diffusivities for O in β-Ti at 1000°C of from 3×10^{-7} to 8×10^{-9} cm^2 s^{-1}. CHARLESWORTH and MADSEN [CHA70] have also assembled a set of diffusivity data (including an *atomic* activation energy expressed in eV) for C, N, O and H in β-Ti as well as in unalloyed Nb at moderate temperatures, *Table 7-4*. In the absence of published diffusivity data for interstitials in Ti-Nb alloys, those authors assumed that an activation energy for diffusion in Ti-Nb(45 at.%), cited in this context as a typical β-isomorphous Ti-TM alloy, could be adequately extracted by linear interpolation between the published values for Nb and β-Ti. The values so obtained are included in *Table 7-4*. Bearing in mind the inaccuracy of diffusivity data for interstitial elements in transition metals, it

Table 7-3. Diffusion Parameters for Oxygen in β-Ti [ZWI74, p. 104]

Temperature range, °C	Frequency factor, D_0, cm^2 s^{-1}	Molar activation energy, kcal mole^{-1}	Diffusivity at 1000°C, D, cm^2 s^{-1}
930-1150	31,400	68	5.0×10^{-8}
932-1142	330	58	3.6×10^{-8}
950-1415	1.6	48	8.4×10^{-9}
907-1030	0.45	36	3.0×10^{-7}
1135-1355	8.3×10^{-2}	31	3.6×10^{-7}

Table 7-4. Diffusion Parameters for the Interstitial Elements C, N, and O, in Nb, β-Ti (at elevated temperatures), and Ti-Nb(45 at.%) [CHA70]

Interstitial element	Solvent	Temperature range, °C	Frequency factor, D_0, cm^2 s^{-1}	Atomic activation energy, eV	Diffusivity at 380°C, D, cm^2 s^{-1}
C	Nb	130-280	4×10^{-3}	1.43	3.7×10^{-14}
		150	1.5×10^{-2}	1.15	2.0×10^{-11}
	β-Ti	950-1150	108	2.1	—
	Ti-Nb(45 at.%)	—	—	1.68-1.80	—
N	Nb	150-300	8.6×10^{-3}	1.51	1.9×10^{-14}
		300	9.8×10^{-2}	1.65	1.8×10^{-14}
	β-Ti	900-1570	3.5×10^{-2}	1.46	—
	Ti-Nb(45 at.%)	—	—	1.49-1.57	—
O	Nb	40-150	2.0×10^{-2}	1.17	1.9×10^{-11}
		150	1.5×10^{-2}	1.2	8.2×10^{-12}
	β-Ti	950-1414	1.6	2.09	—
	Ti-Nb(45 at.%)	—	—	1.68-1.69	—

was suggested by CHARLESWORTH and MADSEN that an atomic activation energy of 1.55 eV for C, N, and O in Ti-Nb(45 at.%) would reasonably well approximate the experimental situation. Finally, a value of 2.0×10^{-2} cm^2 s^{-1} (see the O-entry, *Table 7-4*) was suggested for the frequency factor of O, the most mobile of all three interstitials in Nb. At a temperature of 380°C, for example, these data then yielded a diffusivity for O of 2.2×10^{-14} cm^2 s^{-1}. In order to establish a benchmark for comparison, *Fig. 7-13*, which presents a set of diffusivity data for ^{95}Nb in a series of Ti-Nb alloys (with appropriate extrapolations where necessary), may be consulted. If this is done, it will be found that O diffuses in Ti-Nb(45 at.%) at 380°C as rapidly as Nb does in it at about 800°C. This leads immediately to the interesting conclusion that as cell growth in Ti-Nb begins to accelerate with increase in temperature above 400°C (*Fig. 7-11(b)*), the interstitial atoms will long since have diffused to, and become trapped by, the cell walls. Alternatively, using linear random walk theory, according to which an atom moves a root-mean-square distance from its starting point of $\sqrt{2Dt}$ cm in t seconds, it can be seen that an interstitial atom will migrate a distance of 1300 Å within the Ti-Nb alloy during a heat treatment of 1h/380°C; by the same token, the time taken for it to traverse one-half of a typical subband diameter at that temperature is about $1\frac{1}{2}$ min.

Interstitial-atom diffusion has an important influence on cell-wall precipitation. The elements B, C, N, and O are all stabilizers of the α phase in Ti and as such are expected to encourage α-phase precipitation as they accumulate at the cell boundaries or residual dislocations. Furthermore, as a solution strengthener, dissolved O tends to stabilize the bcc lattice. Taken together, these comments agree with the observation that O retards ω precipitation but accelerates $\beta \rightarrow \alpha + \beta$ decomposition. The accelerating effect of O on the kinetics of α-phase precipitation in Ti-Mo has been graphically demonstrated by DELAZARO and ROSTOCKER [DEL53], *Fig. 7-14*.

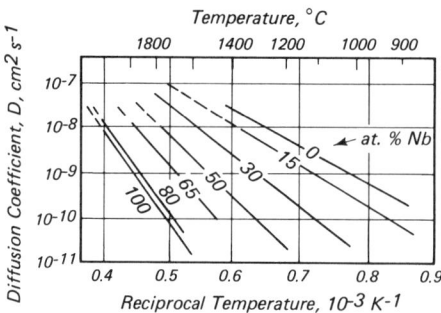

Fig. 7-13. Temperature dependence of the diffusion coefficients of ^{95}Nb in Ti-Nb alloys [ZWI74, p. 112].

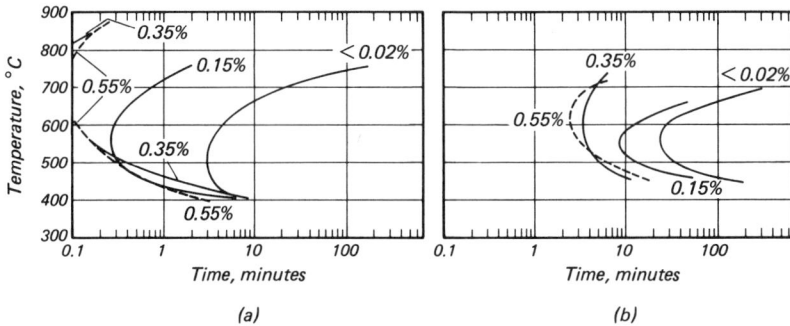

Fig. 7-14. Influence of oxygen (present at the percentages indicated) on the $\beta \rightarrow \alpha + \beta$ transformation kinetics of the β-isomorphous alloy Ti-11Mo; (a) transformation start, (b) transformation finish [DEL 53].

Part 3: The Aging of Some Technical Titanium-Base Alloys

The aging effects encountered in commercial multicomponent alloys, although often complicated by the mixture of phases present in the starting materials depending on their compositions or thermal histories, can be interpreted in terms of the aging properties of binary alloys. As indicated above, further control over the precipitational processes and microstructures can be exerted by associating the heat treatments with mechanical deformation. The results of this so-called "thermomechanical processing" on microstructures and mechanical properties are to be considered elsewhere in this monograph series.

The precipitational effects likely to be encountered during the aging of technical Ti-base alloys are illustrated here by three case studies.

7.12 The Aging of Ti-6Al-4V

The effect of heat treatment on the properties of Ti-6Al-4V, as catalogued by MCQUILLAN [MCQ63, p. 72] and WOOD [WOO72, p. 1-4:72-11], can be appreciated to some extent with the aid of the equilibrium phase diagrams of *Figs. 3-14* through *3-16*. The aging behavior will of course depend on the initial state of the alloy. For example, depending on the temperature at which it was held prior to being quenched, the alloy could consist of either: (i) all martensite, (ii) primary α plus martensite, or (iii) primary α plus retained β [MCQ63]. In case the initial condition is martensitic: (i) the α' recovers towards an α possessing the appropriate equilibrium composition; and (ii) the boundaries and internal structures (Sect.

4.2.2) of the plates provide heterogeneous nucleation sites for the precipitation of β [Wɪʟ73]. The V content of the β and the Al content of the α naturally play important roles in the precipitation process. With regard to the Al component, the precipitation of α_2 can contribute to the strengthening of the alloy.

7.13 The Aging of Ti-6Al-2Sn-4Zr-2Mo

Alloys such as Ti-6Al-2Sn-4Zr-2Mo provide enormous opportunities for studying the effects of thermomechanical processing on microstructure. Again, the aged properties will depend on the combination of initial heat treatment and mechanical deformation to which the material has been subjected. In actual practice, for this alloy the thermomechanical processing will be an integral part of a forging operation. Properties sought are a combination of high strength and fracture toughness and good creep resistance. The creep properties of Ti-6242(Si) alloys can be maximized in the following way [Cʜᴇ80]: (i) for $\alpha + \beta$ forgings, by a β solution treatment* followed by aging for 8h/593°C/air cool; (ii) for β forgings, the solution heat treatment should be conducted in the $\alpha + \beta$ field. The microstructural goal of these processing operations is to achieve a fine Widmanstätten structure and to avoid the occurrence of both martensitic α' and globular α, both of which seem to degrade the creep resistance and the fracture toughness [Cʜᴇ80].

7.14 The Aging of Two Technical β Alloys

7.14.1 Introduction

A useful insight into the aging/mechanical-property characteristics of β-Ti alloys in general can be acquired from a study of the properties of the alloy sequence Ti-15Mo, Ti-15Mo-5Zr, and Ti-15Mo-5Zr-3Al [Pᴇɴ80] [Nɪs82]. The aging of Ti-15Mo and Ti-15Mo-5Zr at temperatures below about 400°C resulted in pronounced ω-phase-induced hardening. In Ti-15Mo-5Zr, the isochronal curves of tensile strength versus temperature and elongation versus temperature were practically mirror-images of each other about a suitably selected horizontal axis; at 400°C, the tensile strengths rose to maxima while the elongations dropped to zero [Nɪs82]. In order to develop α-phase precipitation in Ti-15Mo-5Zr, the aging temperature had to be raised above 400°C. The exchange of ω phase for α phase resulted in a recovery of the ductility. Nɪsʜɪᴍᴜʀᴀ et al. [Nɪs82] found that a satisfactory combination of strength and ductility could be acquired by

---

*The β transus is at 990°C and M_s is at 800°C (see *Fig. 3-19*).

developing a suitable mixture of ω-phase and α-phase precipitates; they did not investigate the ω-reversion phenomenon (see Sect. 7.5). The presence of 3 wt.% Al in the above alloy was sufficient to severely suppress the formation of ω phase. Although such precipitation could be induced to form at temperatures below about 400°C, some two hours of aging were needed to elicit its first appearance. The principal strengthener of Ti-15Mo-5Zr-3Al is α precipitation, which forms at temperatures above about 425°C, and which is relatively rich in Al, a rapid hardener of Ti alloys.

In alloys such as Ti-15Mo, Ti-15Mo-5Zr, and Ti-15Mo-5Zr-3Al, the heat-treatment-induced microstructural mechanisms that control strength and ductility are not amenable to simple explanation. It is not sufficient simply to specify the relative abundances of the ω-phase and α-phase precipitates. Among the other factors which must be taken into consideration are: (i) the grain size of the β matrix [Nis82], and (ii) the size and morphology of the α precipitates — for example, whether they: (a) grow out from the grain boundaries as platelets, (b) exist as a fine distribution of ellipsoidal particles, or (c) assume a coarse lenticular Widmanstätten morphology [Pen80].

7.14.2 The Aging of Ti-11.5Mo-6Zr-4.5Sn (β-III)

As pointed out in Sect. 3.3.4(b), the metallurgical behavior exhibited by β-III is similar to that found in Ti-12Mo [Boy74] which may, therefore, be regarded as its binary prototype. The aging of Ti-Mo(≤ 21 at.%) for several hours at temperatures below 400°C results in the nucleation and growth of isothermal ω-phase precipitates.* In particular, during the aging of Ti-Mo(8 at.%) for 150 h at that temperature, Hickman [Hic69[a]] found that 68% by volume of the alloy had become ω phase with a Mo concentration of 4.7 at.%, while the composition of the β matrix was 14.5 at.% Mo (25.4 wt.%). The presence of ω phase severely embrittles any Ti-base alloy in which it occurs; for this reason an early proclaimed disadvantage of β-III was its unsuitability for service at temperatures between about 200 and 425°C, a very important temperature range within which any high-strength Ti alloy would be expected to perform reliably. It was, however, a simple matter to devise a stabilization heat treatment to avoid this difficulty. In doing so, Froes et al. [Fro73] developed the pseudobinary equilibrium and meta-equilibrium diagram for β-III (Mo variation) depicted in *Fig. 3-24*. The effect of heat treatment on β-III, as discussed by Froes et al. [Fro73], Boyer et al. [Boy74], and Williams et al. [Wil82[a]], can be appreciated with the aid of such a diagram.

---

*As pointed out by Ho and Collings [Ho73[a]] the $\omega/(\omega+\beta)$ and $(\omega+\beta)/\beta$ boundaries of the meta-equilibrium $\omega+\beta$ field at 350°C are at 4.3 and 21 at.% Mo, respectively.

The stabilization treatment suggested by Froes *et al.* [Fro73] was an aging for 8 h in the temperature range 550 ~ 600°C to produce about 40% α phase and a β matrix too rich in Mo to be susceptible to ω-phase embrittlement during subsequent service ($\geq 8,000$ h) at 200 ~ 425°C. The Mo content of the β-III matrix was 17.2 wt.% (9.4 at.%) after the stabilization heat treatment. Polonis and colleagues (e.g., [Luh71], see also [Cha78]) have conducted extensive investigations of the ω-reversion effect with particular reference to the Ti-Cr system in which the kinetics of the $\omega \xrightarrow{500°C} \beta$ reaction are quite rapid (see Sect. 7.5). While extending these studies to β-III, the strengths and ductilities of the variously thermomechanically processed alloy were correlated with the observed microstructures [Boy74]. It was hoped that the success previously obtained with Ti-Cr [Cha72] would be repeated in β-III. In the former case, the $\omega + \beta \xrightarrow{500°C} \beta' + \beta$ reversion was complete after 5 min; the strength of the original ω-strengthened aged alloy was preserved and the ductility was increased. But with β-III, after a similar treatment, the ω phase was only partially reverted. Some of the strengthening arising from the presence of ω phase was retained, and some ductility, as a result of the increased volume fraction of β phase, was acquired. But it was estimated that if complete reversion had been achieved, the strength would have been no greater than that of the solution-treated-and-quenched alloy. Further interesting results were obtained by repeating the aging and reversion heat treatments on alloys which had been deformed after solution heat treatment. In these cases, the α phase which formed at dislocation sites during $\omega + \beta$ aging played a strengthening role at all stages of the heat treatment.

The experiments of Froes *et al.* [Fro73] and Polonis *et al.* [Boy74] were subsequently extended in detail by Williams *et al.* [Wil82a], who studied the microstructural and mechanical responses of β-III to: (i) variation of solution treatment temperature, (ii) variation of $\alpha + \beta$ aging temperature, (iii) duplex aging (ω reversion), and (iv) cold work administered prior to various heat treatments. Some of the results can be summarized with reference, again, to *Fig. 3-24*, and to *Table 7-5*: (i) in alloys $\alpha + \beta$-aged 8h/590°C after solution treatment (ST) and quenching, a change of ST temperature from 790°C (i.e., above $\beta/(\alpha + \beta)$), in which case the α precipitation was uniformly distributed, to 720°C (i.e., below $\beta/(\alpha + \beta)$), which was responsible for a very irregular α-phase precipitate, had no significant effect on the mechanical properties (*Table 7-5*, cf. rows (a) and (b)). (ii) Aging at a lower temperature than 590°C, in particular 8h/480°C, resulted in a closely spaced array of finer α-phase particles and an increase in strength (*Table 7-5*, cf. rows (b) and (c)). (iii) Rows (d) and (e) of *Table 7-5* compare the effects of 8h/510°C aging on samples that had been quenched and quenched-plus-$\omega + \beta$-aged, respectively. In the latter case

Table 7-5. Tensile Properties of Solution Treated and Aged β-III [WLL82[a]]

Row	Condition*	Tensile strength		Yield strength		Elongation, %	Reduction in area, %
		ksi	10^8 N m^{-2}	ksi	10^8 N m^{-2}		
(a)	ST$_1$ + 8h/590°C	142	9.8	133	9.2	12	62
(b)	ST$_2$ + 8h/590°C	140	9.7	133	9.2	11.5	58
(c)	ST$_2$ + 8h/480°C	188	13.0	179	12.3	5	25
(d)	ST$_2$ + 8h/510°C	205	14.1	200	13.8	6.5	39
(e)	ST$_2$ + 50h/370°C + 8h/510°C	204	14.1	197	13.6	4	13

*ST$_1$ = 5min/790°C + water quench. ST$_2$ = 5min/720°C + water quench.

partial ω reversion,* followed in time by α-phase precipitation at the remaining ω-phase sites, seemed to be taking place. As compared to the single aging treatment, the duplex aging, which included a preliminary $\omega + \beta$ aging, engendered a deterioration in the ductility with no improvement in strength.

7.14.3 The Aging of Ti-3Al-8V-5Cr-4Mo-4Zr (β-C)

β-C is a commercial metastable β-Ti alloy characterized by high strength and toughness and hardenability by heat treatment. It resulted from the need for an alloy more β-stabilized than β-III, which was itself less β-stabilized than the Ti-13V-11Cr-3Al alloy which it has replaced (see Sects. 3.3.4(a) and (b)).

As a result, presumably, of its containing the right amounts of Al and Zr, this alloy does not precipitate isothermal ω during aging [WIL71], see also Sect. 7.4. It does, however, undergo phase separation, the β' precipitates eventually giving way to α phase. The morphology of the latter, which is particularly interesting in this system, has been studied in detail by RHODES and PATON [RHO77]. Although those authors have unequivocally identified β' as the precipitate which formed during the early stages of moderate-temperature aging, MORGAN and HAMMOND [MOR80] in a similar study found only disc-shaped "zones". Otherwise, all authors agreed on the nature of the subsequently formed α-phase precipitates.

Following the work of RHODES and PATON [RHO77], it is convenient to treat the aging of the β-quenched (from $\geq 790°C$) alloys in two temperature regimes, 350°C and 500°C: (i) At 350°C, $\beta \to \beta' + \beta$ begins after 24 h and matures after about 7 days. Fine α-phase precipitates begin to develop about this time and agglomerate into "rafts" of particles after about 14 days. Subsequent aging coarsens the rafts. The precipitate is so-called Type-2 α. It is not Burgers related to the β matrix. (ii) At 500°C, a very short-lived β' precipitate is replaced after about 2 h by α-phase plates. There are Type-1 α which, being Burgers related to the β matrix and sharing slip systems with it, are in principle able to be sheared during deformation. They are not stable, however, and decompose internally during the next few hours of aging into Type-2 α [RHO77][MOR80]. As for the mechanical properties, RHODES and PATON [RHO77] found that the best combination of strength and ductility in β-C was achieved in response to the formation at temperatures above 500°C of large non-coherent Type-2 α precipitates.

---

*At a second-stage aging temperature of 540°C, complete reversion of the ω phase was achieved and the hardness dropped to that of the as-ST material, in agreement with the earlier observation of POLONIS et al. [BOY74].

References

A

[AGA74] AGARWAL, K.L., Low Temperature Calorimetry of Transition Metal Alloys, *Ph.D. Thesis*, University of New Orleans, LA, 1974.

[AGE70] AGEEV, N.V. and PETROVA, L.A., The Theoretical Basis of the Development of the High-Strength Metastable β-Alloys of Titanium, in [JAF70], pp. 809-14.

[ALB72] ALBERT, H. and PFEIFFER, I., Uber die Temperaturabhängigkeit des Elastizitätsmoduls von Niob-Titan-Legierungen, *Z. Metallkde.* **63**, 126-31 (1972).

[ALB76] ALBERT, H. and PFEIFFER, I., Temperaturabhängigkeit der Festigkeitseigenschaften des Hochfeldsupraleiters NbTi50, *Z. Metallkde.* **67**, 356-60 (1976).

[ALE67] ALEKSEYEVSKII, N.Ye., IVANOV, O.S., RAYEVSKIY, I.I., and STEPANOV, M.V., Constitution Diagram of the System Niobium-Titanium-Zirconium and Superconducting Properties of the Alloys, *Phys. Met. Metallogr. (USSR)*, **23** No. 1, 28-35 (1967) [Transl. of *Fiz. Met. Metalloved.* **23**, 28-36 (1967)].

[AND63] ANDERSON, O.L., A Simplified Method for Calculating the Debye Temperature from Elastic Constants, *J. Phys. Chem. Solids* **24**, 909-17 (1963).

[ARN74] ARNDT, R. and EBELING, R., Einfluss von Gefügeparametern auf die Stromtragfähigkeit von Niob-Titan-Supraleitern, *Z. Metallkde.* **65**, 364-73 (1974).

[AME54] AMES, S.L. and MCQUILLAN, A.D., The Resistivity-Temperature-Concentration Relationships in the System Niobium-Titanium, *Acta. Metall.* **2**, 831-6 (1954).

B

[BAG58] BAGARIATSKII, Yu.A. and NOSOVA, G.I., A More Accurate Determination of Atomic Coordinates of the Metastable ω-Phase in Ti-Cr Alloys, *Soviet Phys. Cryst.* **3**, 15-26 (1958) [Transl. of *Kristallografiya* **3**, 17-28 (1958)].

[BAG59] BAGARIATSKII, Yu.A., NOSOVA, G.I., and TAGUNOVA, T.V., Factors in the Formation of Metastable Phases in Titanium-Base Alloys, *Sov. Phys. Dokl.* **3**, 1014-8 (1959) [Transl. of *Dok. Akad. Nauk SSSR* **122**, 593-6 (1958)].

[BAK67] BAKER, C. and TAYLOR, M.T., Some Problems of Relating the Critical Current Density to Dislocation Distribution in Worked Superconducting Alloys, *Phil. Mag.* **16**, 1129-32 (1967).

[BAK69] BAKER, C. and SUTTON, J., Correlation of Superconducting and Metallurgical Properties of a Ti-20at.%Nb Alloy, *Phil. Mag.* **19**, 1223-55 (1969).

[BAK70] BAKER, C., The Effect of Heat-Treatment and Nitrogen Addition on the Critical Current Density of a Worked Niobium 44 wt% Titanium Superconducting Alloy, *J. Mater. Sci.* **5**, 40-52 (1970).

[BAK71] BAKER, C., The Shape-Memory Effect in a Titanium-35 wt.% Niobium Alloy, *Metal Science J.* **5**, 92-100 (1971).

[BAL71] BALCERZAK, A.T., An Electron Microscope Study of the Ti-Nb System, *M.S. Thesis*, Cornell University, Ithaca, NY, 1971.

[BAL72] BALCERZAK, A.T. and SASS, S.L., The Formation of the ω Phase in Ti-Nb Alloys, *Met. Trans.* **3**, 1601-5 (1972).

[BAR57] BARDEEN, J., COOPER, L.N., and SCHRIEFFER, J.R., Theory of Superconductivity, *Phys. Rev.* **108**, 1175-201 (1957).

[BAR60] BARTON, J.W., PURDY, G.R., TAGGART, R., and PARR, J.G., Structure and Properties of Titanium-Rich Titanium-Nickel Alloys, *Trans. TMS-AIME* **218**, 844-9 (1960).

[BAS57] BASINSKI, Z.S., The Instability of Plastic Flow of Metals at Very Low Temperatures, *Proc. Roy. Soc.* **A240**, 229-42 (1957).

[BAT64] BATT, R.H., Apparatus for Heat Capacity Measurements, *Ph.D. Thesis*, University of California, Berkeley, CA, 1964.

[BAT73] BATTERMAN, B.W., MARACCI, G., MERLINI, A., and PACE, S., Diffuse Mössbauer Scattering Applied to Dynamics of Phase Transformations, *Phys. Rev. Letters* **31**, 227-30 (1973).

[BIL56] BILBY, B.A. and CHRISTIAN, J.W., Martensitic Transformations, in *The Mechanisms of Phase Transformations in Metals*, The Institute of Metals, 1956, pp. 121-72.

[BLA67] BLANDIN, A.P., Theory of the Hume-Rothery Rules, in *Phase Stability in Metals and Alloys*, ed. by P.S. Rudman, J. Stringer, and R.I. Jaffee, McGraw-Hill, 1967, pp. 115-24.

[BLA67ᵃ] BLACKBURN, M.J., The Ordering Transformation in Titanium:Aluminum Alloys Containing up to 25 at. pct. Aluminum, *Trans. TMS-AIME* **239**, 1200-8 (1967).

[BLA68] BLACKBURN, M.J. and WILLIAMS, J.C., Phase Transformations in Ti-Mo and Ti-V Alloys, *Trans. TMS-AIME,* **242**, 2461-9 (1968).

[BLA69] BLACKBURN, M.J. and WILLIAMS, J.C., Strength, Deformation Modes, and Fracture in Titanium-Aluminum Alloys, *Trans. ASM* **62**, 398-409 (1969).

[BLA70] BLACKBURN, M.J., Some Aspects of Phase Transformations in Titanium Alloys, in [JAF70], pp. 633-43.

[Bog73] BOGATCHEV, I.N. and DYAKOVA, M.A., Titanium Alloys' Behaviour under Various Loading Conditions, in [JAF73], pp. 1119-30.

[Boy74] BOYER, R.R., TAGGART, R., and POLONIS, D.H., Effect of Thermal and Mechanical Processes on the β-III Titanium Alloy, *Metallography* **7**, 241-51 (1974).

[Bra67] BRAMMER, W.G. Jr. and RHODES, C.G., Determination of Omega Phase Morphology in Ti-35%Nb by Transmission Electron Microscopy, *Phil. Mag.* **16**, 477-86 (1967).

[Bre67] BREWER, L., Viewpoints of Stability of Metallic Structures, in *Phase Stability in Metals and Alloys*, ed. by P.S. Rudman, J. Stringer, and R.I. Jaffee, McGraw-Hill, 1967, pp. 39-61.

[Bri70] BRINDLEY, B.J. and WORTHINGTON, P.J. Yield-point Phenomena in Substitutional Alloys, *Metallurgical Reviews* **15**, 101-14 (1970).

[Bro55] BROTZEN, F.R., HARMON, E.L. Jr., and TROIANO, A.R., Decomposition of Beta Titanium, *Trans. TMS-AIME* **203**, 413-9 (1955).

[Bro64] BROWN, A.R.G., CLARK, D., EASTABROOK, J., and JEPSON, K.S., The Titanium-Niobium System, *Nature* **201**, 914-5 (1964).

[Bro65] BROWN, A.R.G., JEPSON, K.S., and HEAVENS, J., High-Speed Quenching in Vacuum, *J. Inst. Metals* **93**, 542-4 (1965).

[Buc65] BUCHER, E., HEINIGER, F., and MÜLLER, J., Anomalies in the Superconducting Transition of HCP Titanium Alloys, in *Low Temperature Physics — LT9* (Proc. 9th Int. Conf., Columbus, OH, Sept. 1964), ed. by J.G. Daunt *et al.*, Plenum Press, 1965, pp. 482-6.

[Buy70] BUYNOV, N.N., VOZILKIN, V.A., and RAKIN, V.G., Structural Study of the Superconductive Alloy 65 BT, *Phys. Met. Metallogr. (USSR)* **29** No. 5, 115-9 (1970) [Trans. of *Fiz. Met. Metalloved.* **29**, 1005-9 (1970)].

C

[Cah61] CAHN, J.W., On Spinoidal Decomposition, *Acta. Met.* **9**, 795-801 (1961).

[Cha70] CHARLESWORTH, J.P. and MADSEN, P.E., Effect of Heat Treatment on the Superconducting Critical Current of Cold-Worked Titanium-45 at.% Niobium: Part I, *Report No. AERE-R-6534*, Atomic Energy Research Establishment, Harwell, UK, Oct. 1970.

[Cha72] CHANDRASEKARAN, V., TAGGART, R., and POLONIS, D.H., Fracture Modes in a Binary Titanium Alloy, *Metallography* **5**, 235-50 (1972).

[Cha72ª] CHANDRASEKARAN, V., TAGGART, R., and POLONIS, D.H., Phase Separation Processes in the Beta Phase of Ti-Mo Binary Alloys, *Metallography* **5**, 393-8 (1972).

[Cha73] CHANDRASEKARAN, V., COTTON, W., NARASIMHAN, S., TAGGART, R., and POLONIS, D.H., Study of Phase Transformation and Superconductivity, *Annual Progr. Report No. RLO-2225-T-13-19*, University of Washington, Seattle, WA, Oct. 1973.

[CHA73ᵃ] CHANDRASEKARAN, V., TAGGART, R., and POLONIS, D.H., Decomposition Processes Prior to Detection of the Omega Phase in Aged Ti-Cr Alloys, *Metallography* **6**, 313-22 (1973).

[CHA74] CHANDRASEKARAN, V., TAGGART, R., and POLONIS, D.H., The Influence of Constitution and Microstructure on the Temperature Coefficient of Resistivity in Ti-Base Alloys, *J. Mater. Sci.* **9**, 961-8 (1974).

[CHA78] CHANDRASEKARAN, V., TAGGART, R., and POLONIS, D.H., An Electron Microscopic Study of the Aged Omega Phase in Ti-Cr Alloys, *Metallography* **11**, 183-98 (1978).

[CHA78ᵃ] CHAKRABORTTY, T.K., MUKHOPADHYAY, T.K., and STARKE, E.A. Jr., The Cyclic Stress-Strain Response of Titanium-Vanadium Alloys, *Acta. Met.* **26**, 909-20 (1978).

[CHE72] CHEN, A.-B., WEISZ, G., and SHER, A., Temperature Dependence of the Electron States and dc Electrical Resistivity of Disordered Binary Alloys, *Phys. Rev.* **5**, 2897-924 (1972).

[CHE80] CHEN, C.C. and COYNE, J.E., Relationships between Microstructure and Mechanical Properties in Ti-6Al-2Sn-4Zr-2Mo~0.1Si Alloy Forgings, in [KIM80], pp. 1197-207.

[CHR65] CHRISTIAN, J.W., Phase Transformations, in *Physical Metallurgy*, ed. by R.W. Cahn, North-Holland, 1965.

[CLA73] CLAPP, P.C., A Localized Soft Mode Theory for Martensitic Transformations, *phys. stat. sol. (b)* **57**, 561-9 (1973).

[COH51] COHEN, M., The Martensitic Transformation, in *Phase Transformations in Solids*, ed. by R. Smoluchowski, J.E. Mayer, and W.A. Weyl, John Wiley & Sons, 1951, pp. 588-660.

[COL58] COLLINGS, E.W. and SMITH, R.D., Measurement of the Principal Magnetic Susceptibility Components of Hexagonal-Close-Packed Crystals of Arbitrary Orientation, *J. Appl. Phys.* **39**, 4462-4 (1968).

[COL69] COLLINGS, E.W. and HO, J.C., Enhancement of Superconducting Transition Temperatures in Martensitic Ti-Mo Alloys, *Phys. Letters* **29A**, 306-7 (1969).

[COL70] COLLINGS, E.W. and HO, J.C., Magnetic Susceptibility and Low Temperature Specific Heat of High-Purity Titanium, *Phys. Rev.* **2**, 235-44 (1970).

[COL70ᵃ] COLLINGS, E.W. and HO, J.C., Physical Properties of Titanium Alloys, in [JAF70], pp. 331-47.

[COL70ᵇ] COLLINGS, E.W. and HO, J.C., Influence of Microstructure on the Superconductivity of a Dilute Ti-Mo Alloy, *Phys. Rev.* **1**, 4289-94 (1970).

[COL71] COLLINGS, E.W. and HO, J.C., Magnetic Susceptibility and Low Temperature Specific Heat of Ti, Zr, and Hf, *Phys. Rev.* **4**, 349-56 (1971).

[COL71ᵃ] COLLINGS, E.W., HO, J.C., and JAFFEE, R.I., Theory of Titanium Alloys for High-Temperature Strength, *Technical Report AFML-TR-71-228*, 1971.

[Col71^b] COLLINGS, E.W., ENDERBY, J.E., and HO, J.C., Influence of Generalized Order-Disorder on the Electron States in Five Classes of Binary Alloy System, in *Electronic Density of States,* Proc. 3rd Materials Research Symposium, Nat. Bur. Stand. (U.S.) Spec. Publ. 323, Dec. 1971, pp. 483-92.

[Col71^c] COLLINGS, E.W., BOYD, J.D., and HO, J.C., Enhancement of the Superconducting Transition Temperature in Deformation-Induced Structures, in *Low Temperature Physics — LT12* (Proc. 12th Int. Conf., Kyoto, Japan, Sept. 1970), ed. by E. Kanda, Academic Press of Japan, 1971, p. 316.

[Col71^d] COLLINGS, E.W. and HO, J.C., Density of States of Transition Metal Binary Alloys in the Electron-to-Atom Ratio Range 4.0 to 6.0, in *Electronic Density of States,* Proc. 3rd Materials Research Symposium, Nat. Bur. Stand. (U.S.) Spec. Publ. 323, Dec. 1971, pp. 587-96.

[Col72] COLLINGS, E.W., HO, J.C., and JAFFEE, R.I., Superconducting Transition Temperature, Lattice Instability, and Electron-to-Atom Ratio of Transition-Metal Binary Solid Solutions, *Phys. Rev.* **5**, 4435-49 (1972).

[Col72^a] COLLINGS, E.W., Experimental Studies of Density-of-States-Related Properties in Ti-Mo Alloys, *Proceedings,* Michigan State University Summer School on Alloys, 1972, pp. 236-69.

[Col73] COLLINGS, E.W., HO, J.C., and JAFFEE, R.I., Physics of Titanium Alloys — II: Fermi Density-of-States Properties and Phase Stability of Ti-Al and Ti-Mo, in [JAF73], pp. 831-42.

[Col73^a] COLLINGS, E.W. and GEGEL, H.L., A Physical Basis for Solid Solution Strengthening and Phase Stability in Alloys of Titanium, *Scripta Met.* **7**, 437-43 (1973).

[Col73^b] COLLINGS, E.W., ENDERBY, J.E., GEGEL, H.L., and HO, J.C., Some Relationships Between the Electronic and Mechanical Properties of Ti Alloys Discussed from the Standpoint of Fundamental Alloy Theory, in [JAF73], pp. 801-14.

[Col74] COLLINGS, E.W., Anomalous Electrical Resistivity, bcc Phase Stability, and Superconductivity in Titanium-Vanadium Alloys, *Phys. Rev.* **9**, 3989-99 (1974).

[Col75] COLLINGS, E.W., and GEGEL, H.L. (eds), *Physics of Solid Solution Strengthening,* Plenum Press, 1975.

[Col75^a] COLLINGS, E.W. and GEGEL, H.L., Physical Principles of Solid Solution Strengthening in Alloys, in [Col75], pp. 147-82.

[Col75^b] COLLINGS, E.W., Magnetic Studies of Omega-Phase Precipitation and Aging in Titanium-Vanadium Alloys, *J. Less-Common Metals* **39**, 63-90 (1975).

[Col75^c] COLLINGS, E.W., HO, J.C., and UPTON, P.E., Low-Temperature Calorimetric Studies of Superconductivity and Microstructure in Titanium-Vanadium Alloys, *J. Less-Common Metals* **42**, 285-301 (1975).

[Col76] COLLINGS, E.W. and HO, J.C., Solute-Induced Lattice Stability as it Relates to Superconductivity in Titanium-Molybdenum Alloys, *Solid State Comm.* **18**, 1493-5 (1976).

[Col78]   COLLINGS, E.W., Anomalous Electrical Resistivity and Magnetic Suscep-tibility Temperature Dependences in Ti-V Alloys Exhibiting Reversible Soft-Phonon-Induced Structural Inhomogeneities, in *Electrical Transport and Optical Properties of Inhomogeneous Media*, ed. by J.C. Garland and D.B. Tanner, AIP Conf. Proc. No. 40, American Institute of Physics, 1978, pp. 410-5.

[Col78ª]   COLLINGS, E.W. and WHITE, J.J., Deformation- and Solute-Induced Microstructural Effects in the Superconductivity of Transition-Metal Alloys, in *Transition Metals 1977*, ed. by M.J.G. Lee, J.M. Perz, and E. Fawcett, The Institute of Physics, Conf. Ser. No. 39, 1978, pp. 645-9.

[Col79]   COLLINGS, E.W., Magnetic Studies of Phase Equilibria in Ti-Al(30-57 at.%) Alloys, *Met. Trans.* **10A**, 463-74 (1979).

[Col80]   COLLINGS, E.W., The Metal Physics of Titanium Alloys: A Review, in [Kim80], pp. 77-132.

[Col81]   COLLINGS, E.W., Titanium Alloy Superconductors, in *Titanium for Energy and Industrial Applications*, ed. by D. Eylon, The Metallurgical Society of AIME, 1981, pp. 143-60.

[Col82]   COLLINGS, E.W., Magnetocrystalline Anisotropy in Monocrystalline and Textured Polycrystalline Ti-Al Alloys, in [Wil82], pp. 1461-73.

[Col82ª]   COLLINGS, E.W., Magnetic Investigations of Electronic Bonding and α through γ Phase Equilibria in the Titanium Aluminum System, in [Wil82], pp. 1391-402.

[Col82ᵇ]   COLLINGS, E.W., Response of the Superconducting Transition Tempera-ture and Other Physical Properties to Phase Transformation, Precipita-tion, and Aging in Titanium-Base Transition-Metal-Binary Alloys, in [Wil82], pp. 1407-19.

[Col84]   COLLINGS, E.W., Previously unpublished data.

[Com67]   COMEY, K.R., The Superconductivity of a Titanium-Niobium Alloy, *M.S. Thesis*, Massachusetts Institute of Technology, Cambridge, MA, 1967.

[Con67]   CONRAD, H., Thermally Activated Deformation of α Titanium Below 0.4 T_M, *Can. J. Phys.* **45**, 581-90 (1967).

[Con70]   CONRAD, H. and JONES, R., Effects of Interstitial Content and Grain Size on the Mechanical Behavior of Alpha Titanium Below 0.4 T_m, in [Jaf70], pp. 489-501.

[Con75]   CONRAD, H., DE MEESTER, B., DONER, M., and OKASAKI, K., Strength-ening of Alpha Titanium by the Interstitial Solutes, in [Col75], pp. 1-45.

[Con81]   CONRAD, H., Effect of Interstitial Solutes on the Strength and Ductility of Titanium, *Progr. Mater. Sci.* **26**, 123-403 (1981).

[Coo73]   COOK, H.E., On the Nature of the Omega Transformation, *Acta. Metall.* **21**, 1445-9 (1973).

[Coo74]   COOK, H.E., A Theory of the Omega Transformation, *Acta. Metall.* **22**, 239-47 (1974).

[Coo75]   COOK, H.E., On First-Order Structural Phase Transitions — I. General Considerations of Pre-Transition and Nucleation Phenomena, *Acta. Metall.* **23**, 1027-39 (1975).

[Coo75ª]   COOK, H.E., On First-Order Structural Phase Transitions — II. The Omega Transformation, *Acta. Metall.* **23**, 1041-54 (1975).

[Cou69]   COURTNEY, T.H. and WULFF, J., Omega Phase Formation in Superconducting Ti Alloys, *Mater. Sci. Eng.,* **4**, 93-7 (1969).

[Cra49]   CRAIGHEAD, C.M., SIMMONS, O.W., MADDEX, P.J., GREENIDGE, C.T., and EASTWOOD, L.W., Preparation and Evaluation of Titanium Alloys, *Summary Report — Part III,* July 30, 1949, to Air Material Command, Wright-Patterson Air Force Base, OH, on Contract No. W 33-038 ac-21229.

D

[Dav75]   DAVIS, L.A., Hardness/Strength Ratio of Metallic Glasses, *Scripta Metall.* **9**, 431-5 (1975).

[Dav79]   DAVIS, R., FLOWER, H.M., and WEST, D.R.F., Martensitic Transformations in Ti-Mo Alloys, *J. Mater. Sci.* **14**, 712-22 (1979).

[Dav79ª]   DAVIS, R., FLOWER, H.M., and WEST, D.R.F., The Decomposition of Ti-Mo Alloy Martensites by Nucleation and Growth and Spinoidal Mechanisms, *Acta. Metall.* **27**, 1041-52 (1979).

[Daw70]   DAWSON, C.W. and SASS, S.L., The As-Quenched Form of the Omega Phase in Zr-Nb Alloys, *Met. Trans.* **1**, 2225-33 (1970).

[Def70]   DE FONTAINE, D., Mechanical Instabilities in the b.c.c. Lattice and the Beta to Omega Phase Transformation, *Acta. Metall.* **18**, 275-9 (1970).

[Def71]   DE FONTAINE, D., PATON, N.E., and WILLIAMS, J.C., The Omega Phase Transformation in Titanium Alloys as an Example of Displacement Controlled Reactions, *Acta. Metall.* **19**, 1153-62 (1971).

[Del52]   DELAZARO, D.J., HANSEN, M., RILEY, R.E., and ROSTOKER, W., Time, Temperature, Transformation Characteristics of Titanium-Molybdenum Alloys, *Trans. TMS-AIME* **194**, 265-9 (1952).

[Del53]   DELAZARO, D.J. and ROSTOKER, W., The Influence of Oxygen Contents on Transformations in a Titanium Alloy Containing 11 Per Cent Molybdenum, *Acta. Metall.* **1**, 676-7 (1953).

[Del74]   DELAEY, L., KRISHNAN, R.V., TAS, H., and WARLIMONT, H., Thermoelasticity, Pseudoelasticity and the Memory Effects Associated with Martensitic Transformations, *J. Mater. Sci.* **9**, 1521-35 (1974).

[Des65]   DESORBO, W., Solute Size and Valence Effect in some Superconducting Alloys of Transition Elements, *Phys. Rev.* **140**, A914-19 (1965).

[Dil65]   DILLAMORE, I.L. and ROBERTS, W.T.., Preferred Orientation in Wrought and Annealed Metals, *Met. Rev.* **10**, 271-380 (1965).

[Doi66]   DOI, T., ISHIDA, H., and UMEZAWA, T., Study of Nb-Zr-Ti Phase Diagram (Studies of Hard Superconductor, II), *Nippon Kinzoku Gakkaishi* **30**, 139-45 (1966).

[Don75] DÖNER, M. and CONRAD, H., Deformation Mechanisms in Commercial Ti-5Al-2.5Sn(0.5 At. Pct O_{eq}) Alloy at Intermediate and High Temperatures (0.3-0.6 T_m), *Met. Trans.* **6A**, 853-61 (1975).

[Due80] DUERIG, T.W., TERLINDE, G.T., and WILLIAMS, J.C., The ω-Phase Reaction in Titanium Alloys, in [Kim80], pp. 1299-305.

[Due80ª] DUERIG, T.W., MIDDLETON, R.M., TERLINDE, G.T., and WILLIAMS, J.C., Stress Assisted Transformations in Ti-10V-2Fe-3Al, in [Kim80], pp. 1503-12.

[Duw53] DUWEZ, P., The Martensitic Transformation Temperature in Titanium Binary Alloys, *Trans. ASM* **45**, 934-40 (1953).

E

[Eas75] EASTON, D.S. and KOCH, C.C., Tensile Properties of Superconducting Composite Conductors and Nb-Ti Alloys at 4.2 K, in [Per75], pp. 431-44.

[Eas77] EASTON, D.S. and KOCH, C.C., Mechanical Properties of Superconducting Nb-Ti Composites, *Adv. Cryo. Eng. (Materials)* **22**, 453-62 (1977).

[Emb66] EMBURY, J.D., KEH, A.S., and FISHER, R.M., Substructural Strengthening in Materials Subject to Large Plastic Strains, *Trans. TMS-AIME* **236**, 1252-60 (1966).

[Eva73] EVANS, D., An Hypothesis Concerning the Training Phenomenon Observed in Superconducting Magnets, *Report No. RL-73-092,* Rutherford High Energy Lab., Chilton, England, Aug. 1973.

F

[Fau82] FAULKNER, J.S., The Modern Theory of Alloys, *Progr. Mater. Sci.* **27**, 1-187 (1982).

[Fed63] FEDOTOV, S.G. and BELOUSOV, O.K., The Elastic Properties of Alloys of Titanium with Molybdenum, Vanadium, and Niobium, *Sov. Phys. Dokl.* **8**, 496-8 (1963) [Transl. of *Dok. Akad. Nauk SSSR* **150**, 77-80 (1963)].

[Fed64] FEDOTOV, S.G. and BELOUSOV, O.K., Elastic Constants of the System Titanium-Niobium, *Phys. Met. Metallogr. (USSR)* **17** No. 5, 83-6 (1964) [Transl. of *Fiz. Met. Metalloved.* **17** No. 5, 732-6 (1964)].

[Fed66] FEDOTOV, S.G., Dependence of the Elastic Properties of Titanium Alloys on Their Composition and Structure, in *Titanium and its Alloys,* ed. by I.I. Kornilov, Akademiya Nauk SSSR (1963); Transl. Israel Program for Scientific Translations Ltd., IPST. Cat. No. 1454 (1966), pp. 199-215.

[Fed73] FEDOTOV, S.G., Peculiarities of Changes in Elastic Properties of Titanium Martensite, in [Jaf73], pp. 871-81.

[Fee70] FEENEY, J.A. and BLACKBURN, M.J., Effect of Microstructure on the Strength, Toughness, and Stress-Corrosion Cracking Susceptibility of a Metastable Beta Titanium Alloy (Ti-11.5Mo-6Zr-4.5Sn), *Met. Trans.* **1**, 3309-23 (1970).

[Fis70]   FISHER, E.S. and DEVER, D., Relation of the C′ Elastic Modulus to Stability of b.c.c. Transition Metals, *Acta. Metall.* **18**, 265-9 (1970).

[Fis70ᵃ]   FISHER, E.S. and DEVER, D., The Single Crystal Elastic Moduli of Beta-Titanium and Titanium-Chromium Alloys, in [JAF70], pp. 373-81.

[Fis75]   FISHER, E.S., A Review of Solute Effects on the Elastic Moduli of bcc Transition Metals, in [COL75], pp. 199-225.

[Fis75ᵃ]   FISHER, E.S., WESTLAKE, D.G., and OCKERS, S.T., Effects of Hydrogen and Oxygen on the Elastic Moduli of Vanadium, Niobium, and Tantalum Single Crystals, *phys. stat. sol. (a)* **28**, 591-602 (1975).

[Fix73]   FIX, D.K., Titanium Precision Forgings, in [JAF73], pp. 441-51.

[Flo73]   FLOWER, H.M., SWANN, P.R., and WEST, D.R.F., Thermally Activated Deformation in Ti 1% Si and Ti 5%Zr 1%Si, in [JAF73], pp. 1143-53.

[Flo82]   FLOWER, H.M., DAVIS, R., and WEST, D.R.F., Martensite Formation and Decomposition in Alloys of Titanium Containing β-Stabilizing Elements, in [WIL82], pp. 1703-15.

[Fri64]   FRIEDEL, J., *Dislocations,* Pergamon Press, 1964 [Transl. of *Les Dislocations*, Gauthier-Villars, 1956].

[Fro54]   FROST, P.D., PARRIS, W.M., HIRSCH, L.L., DOIG, J.R., and SCHWARTZ, C.M., Isothermal Transformation of Titanium-Manganese Alloys, *Trans. ASM* **46**, 1056-74 (1954).

[Fro73]   FROES, F.H., CAPENOS, J.M., and WELLS, M.G.H., Alloy Partitioning in Beta III and Effects on Aging Characteristics, in [JAF73], pp. 1621-33.

G

[Gar72]   GARDE, A.M., SANTHANAM, A.T., and REED-HILL, R.E., The Significance of Dynamic Strain Aging in Titanium, *Acta. Metall.* **20**, 215-20 (1972).

[Geg73]   GEGEL, H.L. and HOCH, M., Thermodynamics of α-Stabilized Ti-X-Y Systems, in [JAF73], pp. 923-31.

[Geg73ᵃ]   GEGEL, H.L., HO, J.C., and COLLINGS, E.W., An Electronic Approach to Solid Solution Strengthening in Titanium Alloys, *Inst. Met. (London) Monogr. Rep. Ser.* **1**, PAP 116, 544-8 (1973).

[Geg80]   GEGEL, H., NADIV, S., and RAJ, R., Dynamic Effects on Flow and Fracture During Isothermal Forging of a Titanium Alloy, *Scripta Metall.* **14**, 241-7 (1980).

[Geg80ᵃ]   GEGEL, H.L., Unpublished research.

[Geh70]   GEHLEN, P.C., The Crystallographic Structure of Ti-Al, in [JAF70], pp. 349-57.

[Gil76]   GILMORE, C.M., FREEDMAN, M., and IMAM, M.A., The Relationship of Axial Strain Induced by Cyclic Torsion to Metal Stability and Fatigue Failure in Ti-6Al-4V, *Eng. Fract. Mech.* **8**, 9-15 (1976).

[Git75]   GITTUS, J., *Creep, Viscoelasticity, and Creep Fracture in Solids,* John Wiley & Sons, 1975.

[GOD73] GODDEN, M.J. and ROBERTS, W.M., Ductility of Ti-Al-Ga Alloys, in [JAF73], pp. 2207-18.

[GOL59] GOLDENSTEIN, A.W., METCALFE, A.G., and ROSTOKER, W., The Effect of Stress on the Eutectoid Decomposition of Titanium-Chromium Alloys, *Trans. ASM* **51**, 1036-52 (1959).

[GOR73] GORYNIN, I.V., CHECHULIN, B.B., USHKOV, S.S., and BELOVA, O.S., A Study of the Nature of the Ductile-Brittle Transition in Beta Titanium Alloys, in [JAF73], pp. 1109-18.

[GRI73] GRIFFITHS, P. and HAMMOND, C., Superplasticity in Large Grained Beta Titanium Alloys, in [JAF73], pp. 1155-67.

[GUL71] GULLBERG, R.B., TAGGART, R., and POLONIS, D.H., On the Decomposition of the Beta Phase in Titanium Alloys, *J. Mater. Sci.* **6**, 384-9 (1971).

[GUS82] GUSEVA, L.N. and DOLINSKAYA, L.K., Metastable Phases in Quenched Titanium Alloys with Transition Elements, in [WIL82], pp. 1559-65.

[GUZ66] GUZEI, L.S., SOKOLOVSKAYA, E.M., and GRIGOR'EV, A.T., Phase Diagram of the Niobium-Titanium System, *Vestnik Moskovskogo Universiteta. Khimiya* **21** No. 5, 79-82 (1966).

H

[HAK61] HAKE, R.R., LESLIE, D.H., and BERLINCOURT, T.G., Electrical Resistivity, Hall Effect and Superconductivity of some b.c.c. Titanium-Molybdenum Alloys, *J. Phys. Chem. Solids* **20**, 177-86 (1961).

[HAK63] HAKE, R.R., LESLIE, D.H., and RHODES, C.G., Giant Anisotropy in the High-Field Critical Currents of Cold-Rolled Transition Metal Alloy Superconductors, in *Low Temperature Physics—LT8* (Proc. 8th Int. Conf., London, 1962), ed. by R.O. Davies, Butterworths, 1963, pp. 342-4.

[HAK64] HAKE, R.R. and CAPE, J.A., Calorimetric Investigation of Localized Magnetic Moments and Superconductivity in Some Alloys of Titanium with Manganese and Cobalt, *Phys. Rev.* **135**, A1151-60 (1964).

[HAL80] HALLAM, P. and HAMMOND, C., The Interface Phase in a Near-α Titanium Alloy, in [KIM80], pp. 1435-41.

[HAM70] HAMMOND, C. and KELLY, P.M., Martensitic Transformations in Titanium Alloys, in [JAF70], pp. 659-76.

[HAM78] HAMMOND, C. and NUTTING, J., The Physical Metallurgy of Superalloys and Titanium Alloys, in *Forging and Properties of Aerospace Materials,* The Metals Society, 1978, pp. 75-102.

[HAN51] HANSEN, M., KAMEN, E.L., KESSLER, H.D., and MCPHERSON, D.J., Systems Titanium-Molybdenum and Titanium-Columbium, *Trans. TMS-AIME* **191**, 881-8 (1951).

[HAN54] HANKE, H., *Prüfung Metallischer Werkstoffe,* VEB Verlag Technik, 1954.

[HAN58] HANSEN, M., *Constitution of Binary Alloys,* Second Ed. (with K. Anderko), McGraw-Hill, 1958.

[HAR56] HARDY, H.K. and HEAL, T.J., Nucleation and Growth Processes in Metals and Alloys, in *The Mechanism of Phase Transformations in Metals,* Inst. of Metals Monograph and Report Ser. No. 18, 1956, pp. 1-46.

[HAT68] HATT, B.A. and RIVLIN, V.G., Phase Transformations in Superconducting Ti-Nb Alloys, *J. Phys. D: Applied Phys.* 1, 1145-9 (1968).

[HAY65] HAYDEN, H.W., MOFFATT, W.G., and WULFF, J., *The Structure and Properties of Materials: Volume III, Mechanical Behavior,* John Wiley & Sons, 1965.

[HEI64] HEINIGER, F. and MÜLLER, J., Bulk Superconductivity in Dilute Hexagonal Titanium Alloys, *Phys. Rev.* 134, 1407-9 (1964).

[HEI74] HEIM, J.R., Superconducting Coil Training and Instabilities Due to the Bauschinger Effect, *Tech. Memo TM-334,* Fermi National Accelerator Lab., 1974.

[HEL71] HELLER, W., Über den Einfluss von Germaniumzusätzen dritter Elemente auf das kritische Magnetfeld von Titan-Niob-Legierungen, *D. Eng. Dissertation,* Universität Erlangen-Nürnberg, West Germany (1971).

[HIC68] HICKMAN, B.S., Precipitation of the Omega Phase in Titanium-Vanadium Alloys, *J. Inst. Metals* 96, 330-7 (1968).

[HIC69] HICKMAN, B.S., The Formation of Omega Phase in Titanium and Zirconium Alloys: A Review, *J. Mater. Sci.* 4, 554-63 (1969).

[HIC69a] HICKMAN, B.S., Omega Phase Precipitation in Alloys of Titanium with Transition Metals, *Trans. TMS-AIME* 245, 1329-35 (1969).

[HID80] HIDA, M., SUKEDAI, E., YOKOHARI, Y., and NAGAKAWA, A., Thermal Instability and Mechanical Properties of Beta Titanium-Molybdenum Alloys, in [KIM80], pp. 1327-34.

[HIL61] HILLERT, M., A Solid-Solution Model for Inhomogeneous Systems, *Acta. Metall.* 9, 525-35 (1961).

[HIL67] HILL, R., *The Mathematical Theory of Plasticity,* Oxford University Press, 1967.

[HIL73] HILLMANN, H., Entwicklung harter Supraleiter, vorzugsweise am Beispiel Nb-Ti Teil I, *Metall* (Berlin) 27, 797-808 (1973).

[HIL73a] HILLMANN, H., Werkstoffe für supraleitende Wechselfeldmagnete mit Ummagnetisierungszeiten der Grössenordnung Sekunde, *Forschungsbericht T 73-03,* Vacuumschmeltze GmbH, Hanau, April 1973.

[HIL76] HILLMANN, H., Werkstoffe für dynamische beanspruchbare supraleitende Magnete, *Forschungsbericht BMFT-FB T 76-13,* Vacuumschmeltze, GmbH, Hanau, June 1976.

[HIL81] HILLMANN, H., Private communication.

[Ho69] HO, J.C., GEHLEN, P.C., COLLINGS, E.W., and JAFFEE, R.I., Phase Stability and Solution Strengthening of Solid Solution Phase Titanium Alloys, *Tech. Rep. AFML-TR-70-1*, Battelle Memorial Institute, Sept. 1969.

[Ho71] HO, J.C. and COLLINGS, E.W., Enhancement of the Superconducting Transition Temperatures of Ti-Mo(5, 7 at.%) Alloys by Mechanical Deformation, *J. Appl. Phys.* **42**, 5144-50 (1971).

[Ho72] HO, J.C. and COLLINGS, E.W., Anomalous Electrical Resistivity in Titanium-Molybdenum Alloys, *Phys. Rev. B* **6**, 3727-38 (1972).

[Ho73] HO, J.C. and COLLINGS, E.W., Calorimetric Studies of Superconductive Proximity Effects in a Two-Phase Ti-Fe(7.5 at.%) Alloy, in *Low Temperature Physics—LT13* (Proc. 13th Int. Conf., Boulder, CO, Aug. 1972), ed. by K.D. Timmerhaus *et al.*, Plenum Press, 1974, pp. 403-7.

[Ho73ᵃ] HO, J.C. and COLLINGS, E.W., Physics of Titanium Alloys—I: Alloying and Microstructural Effects in the Superconductivity of Ti-Mo, in [JAF73], pp. 815-30.

[Hoc71] HOCH, M. and VISWANATHAN, R., Thermodynamics of Titanium Alloys: III, The Ti-Mo System, *Met. Trans.* **2**, 2765-7 (1971).

[Hoc73] HOCH, M., BIRLA, N.C., COLE, S.A., and GEGEL, H.L., The Development of Heat-Resistant Titanium Alloys, *Tech. Report AFML-TR-73-297,* Air Force Materials Lab., Dec. 1973.

[Hoc73ᵃ] HOCH, M., SAKAI, T., KRUPOWICZ, J.J., and DELAHANTY, M., The Titanium-Aluminum-Gallium System, in [JAF73], pp. 935-49.

[Hoc73ᵇ] HOCH, M., Winning and Refining, a Critical Review, in [JAF73], pp. 205-31.

[HOR73] HORIUCHI, T., MONJU, Y., TATARA, I., and NAGAI, N., Phase Transformations of Ti-30Nb-30Zr-7Ta Superconducting Alloy, *Nippon Kinzoku Gakkaishi* **37**, 1057-64 (1973).

[Hou70] HOUGHTON, R.W., SARACHIK, M.P., and KOUVEL, J.S., Anomalous Electrical Resistivity and the Existence of Giant Magnetic Moments in Ni-Cu Alloys, *Phys. Rev. Letters* **25**, 238-9 (1970).

[HUB73] HUBBARD, R.T.J., Low to Intermediate Temperature (Titanium) Alloys: A Critical Review, in [JAF73], pp. 1887-91.

I

[IMG61] IMGRAM, A.G., WILLIAMS, D.N., WOOD, R.A., OGDEN, H.R., and JAFFEE, R.I., Metallurgical and Mechanical Characteristics of High-Purity Titanium-Base Alloys, *WADC Technical Report 59-595, Part II,* March 1961.

[ING69] INGLESFIELD, J.E., Perturbation Theory and Alloying Behaviour, I. Formalism, *J. Phys. C (Solid State Phys.)* **2**, 1285-92 (1969).

J

[JAE62] JAEGER, J.C., *Elastic Fracture and Flow,* Methuen and Co., Second Ed., 1962.

[JAF58] JAFFEE, R.I., The Physical Metallurgy of Titanium Alloys, *Progr. Met. Phys.* **7**, 65-163 (1958).

[JAF70] JAFFEE, R.I. and PROMISEL, N.E. (eds), *The Science Technology and Application of Titanium* (Proc. First Int. Conf. on Titanium, London), Pergamon Press, 1970.

[JAF73] JAFFEE, R.I. and BURTE, H.M. (eds), *Titanium Science and Technology,* (Proc. Second Int. Conf. on Titanium, Boston), Plenum Press, 1973.

[JAF73ᵃ] JAFFEE, R.I., Metallurgical Synthesis; A Critical Review, in [JAF73], pp. 1665-93.

[JAF79] JAFFREY, D., Sources of Acoustic Emission (AE) in Metals — A Review, *Australasian Corr. Eng.,* 1979; Part 1, June, pp. 9-15; Part 2, July/Aug. pp. 9-15; Part 3, Sept./Oct., pp. 25-32.

[JAM70] JAMES, D.W. and MOON, D.M., The Martensitic Transformation in Titanium Binary Alloys and its Effect on Mechanical Properties, in [JAF70], pp. 767-78.

[JAY63] JAYARAMAN, A., KLEMENT, W., and KENNEDY, G.C., Solid-Solution Transitions in Titanium and Zirconium at High Pressures, *Phys. Rev.* **131**, 644-9 (1963).

[JEN65] JENSEN, M.A., MATTHIAS, B.T., and ANDRES, K., Electron Density and Electronic Properties in Noble-Metal Transition Elements, *Science* **150**, 1448-50 (1965).

[JEP70] JEPSON, K.S., BROWN, A.R.G., and GRAY, J.A., The Effect of Cooling Rate on the Beta Transformation in Titanium-Niobium and Titanium-Aluminum Alloys, in [JAF70], pp. 677-90.

K

[KAT79] KATAHARA, K.W., MANGHNANI, M.H., and FISHER, E.S., Pressure Derivatives of the Elastic Moduli of BCC Ti-V-Cr, Nb-Mo, and Ta-W Alloys, *J. Phys. F: Metal Phys.* **9**, 773-90 (1979).

[KAT79ᵃ] KATAHARA, K.W., NIMALENDRAN, M., MANGHNANI, M.H., and FISHER, E.S., Elastic Moduli of Paramagnetic Chromium and Ti-V-Cr Alloys, *J. Phys. F: Metal Phys.* **9**, 2167-76 (1979).

[KAU70] KAUFMAN, L. and BERNSTEIN, H., *Computer Calculations of Phase Diagrams,* Academic Press, 1970.

[KAU73] KAUFMAN, L. and NESOR, H., Phase Stability and Equilibria as Affected by the Physical Properties and Electronic Structure of Titanium Alloys, in [JAF73], pp. 773-800.

[KEA74] KEATING, D.T. and LAPLACA, S.J., Neutron Diffraction Determination of the Number and Displacement of the Atoms in the Diffuse ω-Phase of $Zr_{0.8}Nb_{0.2}$, *J. Phys. Chem. Solids* **35**, 879-91 (1974).

[KEE56] KEELER, J.H. and GEISLER, A.H., Preferred Orientations in Rolled and Annealed Titanium, *Trans. TMS-AIME* **206**, 80-90 (1956).

[KIM80] KIMURA, H. and IZUMI, O. (eds), *Titanium '80: Science and Technology* (Proc. Fourth Int. Conf. on Titanium, Kyoto, Japan), The Metallurgical Society of AIME, 1980.

[KIR70] KIRKPATRICK, S., VELICKÝ, B., and EHRENREICH, H., Paramagnetic Ni-Cu Alloys: Electronic Density of States in the Coherent-Potential Approximation, *Phys. Rev. B* **1**, 3250-63 (1970).

[KIT56] KITTEL, C., *Introduction to Solid State Physics,* Second Ed., John Wiley & Sons, 1956.

[KIT70] KITADA, M. and DOI, T., Discontinuous Precipitation of Solution Treated Nb-40Zr-10Ti Superconducting Alloy, *Nippon Kinzoku Gakkaishi* **34**, 361-5 (1970).

[KIT70a] KITADA, M. and DOI, T., Precipitation and Superconducting Properties of Nb-40Zr-10Ti Alloy, *Nippon Kinzoku Gakkaishi* **34**, 369-74 (1970).

[KOC77] KOCH, C.C. and EASTON, D.S., A Review of Mechanical Behavior and Stress Effects in Hard Superconductors, *Cryogenics* **17**, 391-413 (1977).

[KOR82] KORNILOV, I.I., Equilibrium Diagrams, Electronic and Crystalline Structures and Physical Properties of Titanium Alloys, in [WIL82], pp. 1281-305.

[KOT70] KOT, R., KRAUSE, G., and WEISS, V., Transformation Plasticity of Titanium, in [JAF70], pp. 597-605.

[KOU70] KOUL, M.K. and BREEDIS, J.F., Phase Transformations in Beta Isomorphous Titanium Alloys, *Acta. Metall.* **18**, 579-88 (1970).

[KOU70a] KOUL, M.K. and BREEDIS, J.F., Reply to Comments on "Phase Transformations in Beta Isomorphous Titanium Alloys", *Scripta Metall.* **4**, 877-80 (1970).

[KRA67] KRAMER, D. and RHODES, C.G., Omega Phase Precipitation and Superconducting Critical Transport Currents in Titanium-22 at.% Niobium (Columbium), *Trans. TMS-AIME* **239**, 1612-5 (1967).

[KUB56] KUBASCHEWSKI, O. and CATTERALL, J.A., *Thermochemical Data of Alloys,* Pergamon Press, 1956.

L

[LAR71] LARSON, F.R., ZARKADES, A., and AVERY, D.H., Twinning and Texture Transitions in Titanium Solid-Solution Alloys, *Technical Report 71-11, Accession Number DA 0A4716,* Army Materials and Mechanics Research Center, June 1971.

[LAR74] LARSON, F. and ZARKADES, A., Properties of Textured Titanium Alloys, *Report No. MCIC-74-20,* Metals and Ceramics Information Center, Battelle, June 1974.

[LED80] LEDBETTER, H.M., Sound Velocities and Elastic-Constant Averaging for Polycrystalline Copper, *J. Phys. D: Appl. Phys.* **13**, 1879-84 (1980).

[LED81] LEDBETTER, H.M., Private communication.

[LER60] LERINMAN, R.M., SHCHEGOLEVA, T.V., KUSHAKEVICH, S.A., and SELIT-SKAYA, S.I., Electron Microscope Investigation of the Structural Transformation in Titanium-Manganese and Titanium-Chromium Alloys, *Phys. Met. Metallogr. (USSR)* 9 No. 3, 99 (1960) [Transl. of *Fiz. Met. Metalloved.* 9, 437-40 (1960)].

[LOH71] LOHBERG, R., Uber den Einfluss von Kupferzusätzen auf die supraleitenden Eigenschaften un Phasenumwandlungen von technischen Titan-Niob-Legierungen, *D. Eng. Dissertation*, Universität Erlangen-Nürnberg, West Germany, 1971.

[LOV66] LOVE, G.R. and PICKLESHEIMER, M.L., The Kinetics of Beta-Phase Decomposition in Niobium(Columbium)-Zirconium, *Trans. TMS-AIME* 236, 430-5 (1966).

[LUH68] LUHMAN, T.S., TAGGART, R., and POLONIS, D.H., A Resistance Anomaly in Beta Stabilized Ti-Cr Alloys, *Scripta Metall.* 2, 169-72 (1968).

[LUH69] LUHMAN, T.S., TAGGART, R., and POLONIS, D.H., Correlation of Superconducting Properties with the Beta to Omega Phase Transformation in Ti-Cr Alloys, *Scripta Metall.* 3, 777-83 (1969).

[LUH70] LUHMAN, T.S., Superconductivity and Constitution of Titanium Base Transition Metal Alloys, *Ph.D. Thesis*, University of Washington, Seattle, WA, 1970.

[LUH70[a]] LUHMAN, T.S., TAGGART, R., and POLONIS, D.H., The Effects of Step Quenching and Aging on the Superconducting Transition in Beta Stabilized Ti-Cr Alloys, *Scripta Metall.* 4, 611-5 (1970).

[LUH71] LUHMAN, T.S., TAGGART, R., and POLONIS, D.H., The Effect of Omega Phase Reversion on the Superconducting Transition in Titanium-Base Alloys, *Scripta Metall.* 5, 81-6 (1971).

[LUH72] LUHMAN, T.S., TAGGART, R., and POLONIS, D.H., Magnetic Hysteresis Studies of Superconducting Beta Stabilized Titanium Alloys, *Scripta Metall.* 6, 1055-60 (1972).

[LUK64] LUKE, C.A., TAGGART, R., and POLONIS, D.H., The Metastable Constitution of Quenched Titanium and Zirconium-Base Binary Alloys, *Trans. ASM* 57, 142-9 (1964).

[LUT70] LUTJERING, G. and WEISSMANN, S., Mechanical Properties of Age-Hardened Titanium-Aluminum Alloys, *Acta. Metall.* 18, 785-95 (1970).

[LUT70[a]] LUTJERING, G. and WEISSMANN, S., Mechanical Properties and Structures of Age-Hardened Ti-Cu Alloys, *Met. Trans.* 1, 1641-9 (1970).

[LYE66] LYE, R.G., Band Structure and Bonding in Titanium Carbide, *RIAS Technical Report to NASA*; 2nd Tech. Rept. on Contr. NASw-1290, Nov. 1966.

M

[MAR53] MARGOLIN, H., ENCE, E., and NIELSEN, J.P., Titanium-Nickel Phase Diagram, *Trans. TMS-AIME* 197, 243-7 (1953).

[MAR60] MARGOLIN, H. and NIELSEN, J.P., Titanium Metallurgy, in *Modern Materials, Advances in Development and Application*, ed. by H.H. Hausner, Vol. 2, Academic Press, 1960, pp. 225-325.

[MAR64] MARSH, D.M., Plastic Flow in Glass, *Proc. Roy. Soc.* **A279**, 420-35 (1964).

[MAR64ᵃ] MARSH, D.M., Plastic Flow and Fracture of Glass, *Proc. Roy. Soc.* **A282**, 33-44 (1964).

[MAR77] MARGOLIN, H., LEVINE, E., and YOUNG, M., The Interface Phase in Alpha-Beta Titanium Alloys, *Met. Trans.* **8A**, 373-7 (1977).

[MAR78] MARGOLIN, H., HAZAVEH, F., and YAGUCHI, H., The Grain Boundary Contribution to the Bauschinger Effect, *Scripta Metall.* **12**, 1141-5 (1978).

[MAY53] MAYKUTH, D.J., OGDEN, H.R., and JAFFEE, R.I., Titanium-Tungsten and Titanium-Tantalum Systems, *Trans. TMS-AIME* **197**, 231-7 (1953).

[MAY61] MAYKUTH, D.J., HOLDEN, F.C., WILLIAMS, D.N., OGDEN, H.R., and JAFFEE, R.I., The Effects of Alloying Elements in Titanium: Volume B. Physical and Chemical Properties, Deformation and Transformation Characteristics, *DMIC Report 136B,* Battelle Memorial Institute, May 29, 1961.

[MCC71] MCCABE, K.K. and SASS, S.L., The Initial Stages of the Omega Phase Transformation in Ti-V Alloys, *Phil. Mag.* **23**, 957-70 (1971).

[MCH53] MCHARGUE, C.J., ADAIR, S.E., and HAMMOND, J.P., Effects of Solid Solution Alloying on the Cold-Rolled Texture of Titanium, *Trans. TMS-AIME* **197**, 1199-203 (1953).

[MCQ56] MCQUILLAN, A.D. and MCQUILLAN, M.K., *Titanium,* Academic Press, 1956.

[MCQ63] MCQUILLAN, M.K., Phase Transformations in Titanium and its Alloys, *Metallurgical Reviews* **8**, 41-104 (1963).

[MEA65] MEADEN, G.T., *Electrical Resistance of Metals,* Plenum Press, 1965.

[MEN71] MENDIRATTA, M.G., LUTJERING, G., and WEISSMANN, S., Strength Increase in Ti 35Wt Pct Nb Through Step-Aging, *Met. Trans.* **2**, 2599-605 (1971).

[MOL65] MOLCHANOVA, E.K., *Phase Diagrams of Titanium Alloys* [Transl. of *Atlas Diagram Sostoyaniya Titanovyk Splavov*], Israel Program for Scientific Translations, Jerusalem, 1965.

[MOO73] MOOIJ, J.H., Electrical Conduction in Concentrated Disordered Transition Metal Alloys, *phys. stat. sol. (a)* **17**, 521-30 (1973).

[MOR80] MORGAN, C.C. and HAMMOND, C., The Ageing Characteristics of Ti-3%Al-8%V-6%Cr-4%Mo-4%Zr (Ti-38644), in [KIM80], pp. 1443-51.

[MOS73] Moss, S.C., KEATING, D.T., and AXE, J.D., Neutron Study of the Beta-to-Omega Instability in $Zr_{0.80}Nb_{0.20}$, in *Phase Transitions, 1973*, ed. by L.E. Cross, Pergamon Press, 1973, pp. 179-88.

[Mos80] MOSKALENKO, V.A., STARTSEV, V.I., and KOVALEVA, V.N., Low Temperature Peculiarities of Plastic Deformation in Titanium and its Alloys, in [KIM80], pp. 821-30.

[MYR75] MYRON, H.W., FREEMAN, A.J., and MOSS, S.C., Electronically Induced Lattice Instabilities in bcc Zr, *Solid State Comm.* **17**, 1467-70 (1975).

N

[NAR66] NARLIKAR, A.V. and DEW-HUGHES, D., Superconductivity in Deformed Niobium Alloys, *J. Mater. Sci.* **1**, 317-35 (1966).

[NAR70] NARAYANAN, G.H. and ARCHBOLD, T.F., Comments on "Phase Transformations in Beta Isomorphous Titanium Alloys", *Scripta Metall.* **4**, 873-6 (1970).

[NAR71] NARAYANAN, G.H., LUHMAN, T.S., ARCHBOLD, T.F., TAGGART, F., and POLONIS, D.H., A Phase Separation Reaction in a Binary Titanium-Chromium Alloy, *Metallography* **4**, 343-58 (1971).

[NEA71] NEAL, D.F., BARBER, A.C., WOOLCOCK, A., and GIDLEY, J.A.F., Structure and Superconducting Properties of Nb 44 Percent Ti Wire, *Acta. Metall.* **19**, 143-9 (1971).

[NIS76] NISHIHARA, T. and IGUCHI, N., The Shape-Memory Effect under Transformation Superplasticity of Ti-6Al-4V, *Nippon Kinzoku Gakkaishi* **40**, 51-6 (1976).

[NIS82] NISHIMURA, T., NISHIGAKI, M., and KUSAMICHI, H., Aging Characteristics of Beta Titanium Alloys, in [WIL82], pp. 1675-89.

O

[OBS80] OBST, B., PATTANAYAK, D., and HOCHSTUHL, P., Structural Effects in the Superconductor NbTi65, *J. Low-Temp. Phys.* **41**, 595-609 (1980).

[OHT73] OHTANI, S., NISHIGAKI, M., and NISHIMURA, T., The Characteristics of Ti-Mo Beta Titanium Alloy, in [JAF73], pp. 1945-56.

[OKA73] OKAZAKI, K., MOMOCHI, M., and CONRAD, H., Thermally Activated Deformation of Ti-N Alloys, in [JAF73], pp. 1131-42.

[OKA80] OKA, M. and TANIGUCHI, Y., Crystallography of Stress-Induced Products in Metastable Beta Ti-Mo Alloys, in [KIM80], pp. 709-15.

[ORR55] ORRELL, F.R. and FONTANA, M.G., The Titanium-Cobalt System, *Trans. ASM* **47**, 554-64 (1955).

[OSA80] OSAMURA, K., MATSUBARA, E., MIYATANI, T., MURUKAMI, Y., HORIUCHI, T., and MONJU, Y., Effect of Cold Working on Precipitation Behaviour in Superconducting Ti-Nb Alloys, *Phil. Mag. A* **42**, 575-89 (1980).

[OTS61] OTS PB 171424, *DMIC Report 136 B*, Battelle Memorial Institute, May 29, 1961, p. 106.

[OTT70] OTTE, H.M., Mechanism of the Martensitic Transformation in Titanium and its Alloys, in [JAF70], pp. 645-57.

P

[PAR53] PARRIS, W.M., HIRSCH, L.L., and FROST, P.D., Low Temperature Aging in Titanium Alloys, *Trans. TMS-AIME* **197**, 178-9 (1953).

[PAR73] PARRIS, W.M. and RUSSELL, H.A., A New Titanium Alloy for Elevated Temperature Application, in [JAF73], pp. 2219-25.

[PAS78] PASZTOR, G. and SCHMIDT, C., Dynamic Stress Effects in Technical Superconductors and the "Training" Problem of Superconducting Magnets, *J. Appl. Phys.* **49**, 886-99 (1978).

[PAU56] PAULING, L., The Electronic Structures of Metals and Alloys, in *Theory of Alloy Phases,* American Society for Metals, 1956, pp. 220-42.

[PAU67] PAULING, L., *The Chemical Bond,* Cornell University Press, 1967, Chap. 11.

[PEN80] PENNOCK, G.M., FLOWER, H.M., and WEST, D.R.F., The Control of α Precipitation by Two Step Ageing in β Ti-15%Mo, in [KIM80], pp. 1343-51.

[PER75] PERKINS, J. (ed.), *Shape-Memory Effect in Alloys,* Plenum Press, 1975.

[PER75a] PERKINS, J., EDWARDS, G.R., SUCH, C.R., JOHNSON, J.M., and ALLEN, R.R., Thermomechanical Characteristics of Alloys Exhibiting Martensitic Thermoelasticity, in [PER75], pp. 273-99.

[PET71] PETERSON, V.C. and BUEHL, R.C., Methods for Melting Titanium-Base Alloy, *U.S. Patent No. 3,552,947,* Jan. 5, 1971.

[PET72] PETTIFOR, D.G., Theory of the Crystal Structures of Transition Metals at Absolute Zero, in *Metallurgical Chemistry,* ed. by O. Kubaschewski, National Physical Laboratory, HMSO, 1972, pp. 191-9.

[PET73] PETERSON, V.C., FROES, F.H., and MALONE, R.F., Metallurgical Characteristics and Mechanical Properties of Beta III, A Heat-Treatable Titanium Alloy, in [JAF73], pp. 1969-80.

[PET79] PETTIFOR, D.G., Theory of the Heats of Formation of Transition-Metal Alloys, *Phys. Rev. Letters* **42**, 846-50 (1979).

[PFE68] PFEIFFER, I. and HILLMANN, H., Der Einfluss der Struktur auf die Supraleitungseigenschaften von NbTi50 und NbTi65, *Acta. Metall.* **16**, 1429-39 (1968).

[POL55] POLONIS, D.H. and PARR, J.G., Martensite Formation in Powders and Lump Specimens of Ti-Fe Alloys, *Trans. TMS-AIME* **203**, 64 (1955).

[POL69] POLONIS, D.H., A Study of Phase Transformations and Superconductivity, *Progress Report No. 10* and *Report No. RLO-1375-18,* University of Washington, Seattle, WA, Oct. 1969.

[POL70] POLONIS, D.H., A Study of Phase Transformations and Superconductivity, *Annual Progress Report No. RLO-2225-T-13-6,* University of Washington, Seattle, WA, Oct. 1970.

[POL71] POLONIS, D.H., A Study of Phase Transformations and Superconductivity, *Annual Progress Report No. RLO-2225-T-13-9, N72-24786,* University of Washington, Seattle, WA, Oct. 1971.

[Pos81]   Postans, P.J. and Jeal, R.H., Titanium for Fuel Efficient Gas Turbines, in *Titanium for Energy and Industrial Applications,* ed. by D. Eylon, The Metallurgical Society of AIME, 1981, pp. 183-97.

[Pre74]   Prekul, A.E., Rassokhin, V.A., and Volkenshtein, N.V., Effect of Spin Fluctuations on the Superconducting and Normal Properties of Ti Containing V, Nb, or Ta, *Sov. Phys. JETP* **40**, 1134-6 (1974) [Transl. of *Zh. Eksp. Teor. Fiz.* **67**, 2286-92 (1974)].

[Pre76]   Prekul, A.F., Shcherbakov, A.S., and Volkenshtein, N.V., Resistivity and Anomalous Superconducting Transition in $Ti_{1-x}Fe_x$ Alloys ($0 < x \leq 0.2$), *Sov. J. Low Temp. Phys.* **2**, 684-6 (1976) [Transl. of *Fiz. Nizk Temp.* **2**, 1399-404 (1976)].

R

[Rac70]   Rack, H.J., Kalish, D., and Fike, K.D., Stability of As-Quenched Beta-III Titanium Alloy, *Mater. Sci. Eng.* **6**, 181-98 (1970).

[Rac75]   Rack, H.J., Dynamic Strain Aging of Metastable Beta Titanium Alloys, *Scripta Metall.* **9**, 829-31 (1975).

[Ras72]   Rassmann, G. and Illgen, L., Zum Zusammenhang zwischen Gefüge und kritischer Stromdichte bei supraleitenden binären Titan-Niob-Legierungen, *Neue Hütte* **17**, 321-8 (1972).

[Rau56]   Rausch, J.J., Crossley, F.A., and Kessler, H.D., Titanium-Rich Corner of the Ti-Al-V System, *Trans. AIME, Journal of Metals* **8**, 211-4 (1956).

[Rea78]   Read, D.T., Metallurgical Effects in Niobium-Titanium Alloys, *Cryogenics* **18**, 579-84 (1978).

[Ree77]   Reed, R.P., Mikesell, R.P., and Clark, A.F., Low Temperature Tensile Behavior of Copper-Stabilized Niobium-Titanium Superconducting Wire, *Adv. Cryo. Eng. (Materials)* **22**, 463-71 (1977).

[Rei73]   Reid, C.N., Routbort, J.L., and Maynard, R.A., Elastic Constants of Ti-40 at.% Nb at 298 K, *J. Appl. Phys.* **44**, 1398-9 (1973).

[Reu66]   Reuter, F.E., Ralls, K.M., and Wulff, J., Microstructure and Superconductivity of a 44.7 At.Pct Niobium (Columbium)-54.3 At.Pct Titanium Alloy Containing Oxygen, *Trans. TMS-AIME* **236**, 1143-51 (1966).

[Rez82]   Reznichenko, V.A., Physiochemical Principles and Research Trends in New Methods of Titanium Production, in [Wil82], pp. 63-77.

[Rez82a]   Reznichenko, V.A., Karyasin, I.A., Rogatin, A.A., Kipricj, N.A., Kashkarov, A.Z., Zhachkin, V.N., Menyailova, G.A., and Denisov, S.I., Preparation of High-Titanium Slags, in [Wil82], pp. 79-100.

[Rho75]   Rhodes, C.G. and Williams, J.C., Observations of an Interface Phase in the α/β Boundaries in Titanium Alloys, *Met. Trans.* **6A**, 1670-1 (1975).

[Rho77]   Rhodes, C.G. and Paton, N.E., The Influence of Microstructure on Mechanical Properties in Ti-3Al-8V-6Cr-4Mo-4Zr (Beta-C), *Met. Trans.* **8A**, 1749-61 (1977).

[Riz74] Rizzuto, C., Formation of Localized Moments in Metals: Experimental Bulk Properties, *Rep. Progr. Phys.* **37**, 147-229 (1974).

[Rol71] Rolinski, E.J., Hoch, M., and Oblinger, C.J., Determination of Thermodynamic Interaction Parameters in Solid V-Ti Alloys Using the Mass Spectrometer, *Met. Trans.* **2**, 2613-8 (1971).

[Rol72] Rolinski, E.J., Hoch, M., and Oblinger, C.J., Determination of Thermodynamic Interaction Parameters in Solid V-Ti-Cr Alloys Using the Mass Spectrometer, *Met. Trans.* **3**, 1413-8 (1972).

[Rom71] Romero, C.J., The Effect of Microstructure and Preferred Orientation on the Mechanical Behavior of Titanium Alloy Forgings, *LR 24347,* Lockheed-California Co., June 1971.

[Ron70] Ronami, G.N., Kuznetsova, S.M., Fedotov, S.G., and Konstantinov, K.H., Determination of Phase Boundaries of Ti with V, Nb, Mo by using the Method of Diffusing Layers, *Vestnik Moskovskogo Univ. Fiz. Astron. (U.S.S.R.),* No. 2, 186-9 (1970) [in Russian].

[Ros73] Rosenberg, H.W., High Temperature Alloys: A Critical Review, in [Jaf73], pp. 2127-40.

S

[Sak69] Sakai, T., Study of the Titanium-Rich Region of the Titanium-Aluminum-Gallium Ternary System, *M.S. Thesis,* University of Cincinnati, Aug. 1969.

[Sal79] Salmon, D.R., *Low Temperature Data Handbook, Titanium and Titanium Alloys,* National Physical Laboratory, NPL Report QU53 (N 80 23448), May 1979.

[Sal79a] Saleh, Y. and Margolin, H., Bauschinger Effect During Cyclic Straining of Two Ductile Phase Alloys, *Acta. Metall.* **27**, 535-44 (1979).

[Sar70] Sargent, G.A. and Conrad, H., Stress Relaxation and Thermally Activated Deformation in a Titanium-4 wt% Aluminum Alloy, *Scripta Metall.* **4**, 129-33 (1970).

[Sar72] Sargent, G.A. and Conrad, H., On the Strengthening of Titanium by Oxygen, *Scripta Metall.* **6**, 1099-101 (1972).

[Sas69] Sass, S.L., The ω Phase in a Zr-25 at.%Ti Alloy, *Acta. Metall.* **17**, 813-20 (1969).

[Sas72] Sass, S.L., The Structure and Decomposition of Zr and Ti b.c.c. Solid Solutions, *J. Less-Common Metals* **28**, 157-73 (1972).

[Sas75] Sasano, H. and Kimura, H., Serrated Yielding in Ti-2 at.% Zr Alloy, *Nippon Kinzoku Gakkaishi* **39**, 142-7 (1975).

[Sas82] Sasano, H. and Kimura, H., Serrated Flow in α-Titanium Alloys, in [Wil82], pp. 539-51.

[Sch76] Schmidt, C., Investigation of the Training Problem of Superconducting Magnets, *Appl. Phys. Letters* **28**, 463-5 (1976).

[SCH77]   SCHMIDT, C., Effect of Dynamic Stress on Commercial Superconductors: A Test Facility, *Rev. Sci. Instrum.* **48**, 597-601 (1977).

[SCH77a]   SCHMIDT, C., Superconductors under Dynamic Mechanical Stress, *IEEE Trans. Magn.* **MAG-13**, 116-9 (1977).

[SHI77]   SHIBATA, M. and ONO, K., On the Minimization of Strain Energy in the Martensitic Transformation of Titanium, *Acta Metall.* **25**, 35-42 (1977).

[SHU69]   SHUNK, F.A., *Constitution of Binary Alloys, Second Supplement,* McGraw-Hill, 1969.

[SIL58]   SILCOCK, J.M., An X-Ray Examination of the ω Phase in TiV, TiMo, and TiCr Alloys, *Acta Metall.* **6**, 481-92 (1958).

[SIN75]   SINHA, S.K. and HARMON, B.N., Electronically Driven Lattice Instabilities, *Phys. Rev. Letters* **35**, 1515-18 (1975).

[SIN76]   SINHA, S.K. and HARMON, B.N., Phonon Anomalies in d-Band Metals and their Relationship to Superconductivity, in *Superconductivity in d- and f-Band Metals,* ed. by D.H. Douglass, Plenum Press, 1976, pp. 269-96.

[SMI76]   SMITH, H.G., WAKABAYASHI, N., and MOSTOLLER, M., Phonon Anomalies in Transition Metals, Alloys and Compounds, in *Superconductivity in d- and f-Band Metals,* ed. by D.H. Douglass, Plenum Press, 1976, pp. 223-49.

[SOE69]   SOENO, K. and KURODA, T., Kinetics of Beta-Phase Decomposition and the Precipitation of Alpha-Zirconium in Nb-40at.%Zr-10at.%Ti Superconducting Alloy, *Nippon Kinzoku Gakkaishi* **33**, 791-5 (1969).

[SPA57]   SPARKS, C.J. Jr., McHARGUE, C.J., and HAMMOND, J.P., Effects of Aluminum on the Cold-Rolled Textures of Titanium, *Trans. TMS-AIME* **209**, 49-50 (1957).

[SPA58]   SPACHNER, S.A., Comparison of Structure of Omega Transition Phase in Three Titanium Alloys, *Trans. TMS-AIME* **212**, 57-9 (1958).

[STE68]   STERN, E.A., Requirements for a Theory of Disordered Alloys, in *Energy Bands in Metals and Alloys,* ed. by L.H. Bennett and J.T. Waber, Metallurgical Society Conferences Volume 45, Gordon and Breach, 1968, pp. 151-73 (see also *Phys. Rev.* **144**, 545 (1966); *Physics* **1**, 255 (1965)).

[STE75]   STERN, E.A., Application of Alloy Physics to Solution Strengthening, in [COL75], pp. 183-97.

[STE76]   STEELE, R.K. and McEVILY, A.J., The High-Cycle Fatigue Behavior of Ti-6Al-4V Alloy, *Eng. Fracture Mech.* **8**, 31-7 (1976).

[STO78]   STOCKS, G.M., TEMMERMAN, W.M., and GYORFFY, B.L., Complete Solution of the Korringa-Kohn-Rostoker Coherent-Potential Approximation Equations: Cu-Ni Alloys, *Phys. Rev. Letters* **41**, 339-41 (1978).

[STR68]   STRONGIN, M., KAMMERER, O.F., CROW, J.E., PARKS, R.D., DOUGLASS, D.H. Jr., and JENSEN, MA., Enhanced Superconductivity in Layered Metallic Films, *Phys. Rev. Letters* **21**, 1320-1 (1968).

[STR82]   *Structural Alloys Handbook:* 1982 Supplement, Produced and Published by Battelle's Columbus Laboratories, Columbus, OH.

[SUD68]   SUDEREVA, S.V., BUYNOV, N.N., and RAKIN, V.G., Electron Microscopic and X-ray Diffraction Analysis of the Quenched Alloy Ti-25 at.%Nb, *Phys. Met. Metallogr. (USSR)* **26** No. 5, 14-20 (1968) [Transl. of *Fiz. Met. Metalloved.* **26**, 781-8 (1968)].

[SUZ75]   SUZUKI, T. and WUTTIG, M., Analogy Between Spinodal Decomposition and Martensitic Transformation, *Acta Metall.* **23**, 1069-76 (1975).

[SWA58]   SWANN, P.R. and PARR, J.G., Phase Transformations in Titanium-Rich Alloys of Titanium and Cobalt, *Trans. TMS-AIME* **212**, 276-9 (1958).

[SWA73]   SWANSON, M.L. and QUENNEVILLE, A.F., The Effect of Compressional Plastic Deformation on the Superconducting Transition Temperature of Tin Alloys, *Scripta Metall.* **7**, 1011-7 (1973).

T

[TER80]   TERLINDE, G.T., DUERIG, T.W., and WILLIAMS, J.C., The Effect of Heat Treatment on Microstructure and Tensile Properties of Ti-10V-2Fe-3Al, in [KIM80], pp. 1571-81.

[TER82]   TERAUCHI, S., MATSUMOTO, H., SUGIMOTO, T., and KAMEI, K., Investigation of the Titanium-Molybdenum Binary Phase Diagram, in [WIL82], pp. 1335-49.

[THE82]   THEODORSKI, G. and KOSS, D.A., The Cyclic Stress-Strain Response of Age-Hardenable Beta Titanium Alloys, in [WIL82], pp. 553-67.

[THO73]   THORNBURG, D.R. and PIEHLER, H.R., Cold-Rolling Texture Development in Titanium and Titanium-Aluminum Alloys, in [JAF73], pp. 1187-97.

[TIM56]   TIMOSHENKO, S., *Strength of Materials, Part II: Advanced Theory and Problems*, Third Ed., Robert E. Kreiger Publishing Co., 1956.

[TON75]   TONG, H.C. and WAYMAN, C.M., Thermodynamic Considerations of "Solid State Engines" Based on Thermoelastic Martensitic Transformations and the Shape Memory Effect, *Met. Trans.* **6A**, 29-32 (1975).

[TOR80]   TORAN, J.R. and BIEDERMAN, R.R., Phase Transformation Study of Ti-10V-2Fe-3Al, in [KIM80], pp. 1491-501.

[TRE82]   TRENOGINA, T.L. and LERINMAN, R.M., Decomposition of the Martensite in Two-Phase Titanium Alloys, in [WIL82], pp. 1623-32.

[TWE64]   TWEEDALE, J.G., *The Mechanical Properties of Metals, Assessment and Significance,* American Elsevier Publishing Co., 1964.

[TYS75]   TYSON, W.R., Solution Hardening by Interstitials in Close-Packed Metals, in [COL75], pp. 47-77.

V

[VAN64]   VAN OSTENBURG, D.O., LAM, D.J., TRAPP, H.D., PRACHT, D.W., and ROWLAND, T.J., Nuclear Magnetic Resonance and Magnetic Susceptibilities of Alloys of V with Al, *Phys. Rev.* **135-A**, 455-9 (1964).

[VET68] VETRANO, J.B., GUTHRIE, G.L., KISSINGER, H.E., BRIMHALL, J.L., and MASTEL, B., Superconductivity Critical Current Densities in Ti-V Alloys, *J. Appl. Phys.* **39**, 2524-8 (1968).

[VIG82] VIGIER, G., MERLIN, J., and GOBIN, P.F., Decomposition of the Solid Solution in the All-Beta βIII, in [WIL82], pp. 1691-701.

[VOD80] VODOLAZSKY, V.P., KATAJA, V.K., ALEKSANDROV, V.K., KAGANOVICH, A.Z., VOLKOV, V.A., and NEFED'EV, E.I., Study of Titanium Alloys' Deformation Process by Acoustic Emission, in [KIM80], pp. 841-8.

W

[WES80] WEST, A.W. and LARBALESTIER, D.C., Transmission Electron Microscopy of Commercial Filamentary Nb-Ti Superconducting Composites, *Adv. Cryo. Eng. (Materials)* **26**, 471-8 (1980).

[WES82] WEST, A.W. and LARBALESTIER, D.C., α-Ti Precipitation in Niobium-Titanium Alloys, *Adv. Cryo. Eng. (Materials)* **28**, 337-44 (1982).

[WES83] WEST, A.W. and LARBALESTIER, D.C., α-Ti Precipitates in High Current Density Multifilamentary Niobium Titanium Composites, *IEEE Trans. Magn.* **MAG-19**, 548-51 (1983).

[WHI76] WHITE, J.J. and COLLINGS, E.W., Analysis of Calorimetrically Observed Superconducting Transition Temperature Enhancement in Ti-Mo(5 at.%)-Based Alloys, *Magnetism and Magnetic Materials—1976* (Joint MMM-Intermag Conference, Pittsburgh), AIP Conference Proceedings No. 34, pp. 75-7.

[WIL69] WILLIAMS, J.C. and BLACKBURN, M.J., The Structure, Mechanical Properties and Deformation Behavior of Ti-Al and Ti-Al-X Alloys, *Proceedings of the Third Bolton Landing Conference,* Aug. 27, 1969.

[WIL71] WILLIAMS, J.C., HICKMAN, B.S., and LESLIE, D.H., The Effect of Ternary Additions on the Decomposition of Metastable β-Phase Titanium Alloys, *Met. Trans* **2**, 477-84 (1971).

[WIL73] WILLIAMS, J.C., Kinetics and Phase Transformations: A Critical Review, in [JAF73], pp. 1433-94.

[WIL75] WILLBRAND, J. and SCHLUMP, W., Einfluss von Ausscheidungsdichte und Teilchiengrösse auf die Stromtragfähigkeit von NbTi-Supraleitern, *Z. Metallkde.* **66**, 714-9 (1975).

[WIL75ᵃ] WILLBRAND, J., ARNDT, R., EBELING, R., and MOHS, R., Optimierung supraleitender NbTi Legierungen, *Forschungsbericht T 75-35*, Fried. Krupp GmbH, Krupp Forschungsinstitut, Essen, Nov. 1975.

[WIL76] WILLIAMS, J.C., Precipitation in Titanium-Base Alloys, in *Precipitation Processes in Solids,* ed. by K.C. Russell and H.I. Aaronson, The Metallurgical Society of AIME, 1978, pp. 191-224.

[WIL82] WILLIAMS, J.C. and BELOV, A.F. (eds), *Titanium and Titanium Alloys, Scientific and Technological Aspects* (Proc. Third Int. Conf. on Titanium, Moscow), Plenum Press, 1982.

[WIL82^a] WILLIAMS, J.C., FROES, F.H., and FUJISHIRO, S., Microstructure and Properties of the Alloy Ti-11.5Mo-6Zr-4.5Sn (Beta III), in [WIL82], pp. 1421-36.

[WIL82^b] WILLIAMS, J.C., Phase Transformations in Ti Alloys—A Review of Recent Developments, in [WIL82], pp. 1477-98.

[WIN73] WINSTONE, M.R., RAWLINGS, R.D., and WEST, D.R.F., Dynamic Strain Aging in Some Titanium-Silicon Alloys, *J. Less-Common Metals* 31, 143-50 (1973).

[WIT73] WITCOMB, M.J. and DEW-HUGHES, D., Superconductivity of Heat-Treated Nb-65at.%Ti Alloy, *J. Mater. Sci.* 8, 1383-400 (1973).

[WOO72] WOOD, R.A., *Titanium Alloys Handbook*, Metals and Ceramics Information Center, Battelle, Publication No. MCIC-HB-02, Dec. 1972.

Y

[YAK61] YAKYMYSHYN, F.W., PURDY, G.R., TAGGART, R., and PARR, J.G., The Relationship Between the Constitution and Mechanical Properties of Titanium-Rich Alloys of Titanium and Cobalt, *Trans. ASM* 53, 283-94 (1961).

[YAO61] YAO, Y.L., Magnetic Susceptibilities of Titanium-Rich Titanium-Aluminum Alloys, *Trans. ASM* 54, 241-6 (1961).

[YOS56] YOSHIDA, S. and TSUYA, Y., The Temperature Dependence of the Electrical Resistivity of the β-Phase Titanium-Molybdenum Alloys, *J. Phys. Soc. Japan* 11, 1206-7 (1956).

Z

[ZEN48] ZENER, C., *Elasticity and Anelasticity of Metals*, University of Chicago Press, 1948.

[ZEY71] ZEYFANG, R. and CONRAD, H., Deformation Dynamics of a B.C.C. Titanium Alloy (15.2 at.%Mo) Below 650 K ($\sim$0.4 T_m), *Acta. Metall.* 19, 985-90 (1971).

[ZUB79] ZUBECK, R.B., BARBEE, T.W. Jr., GEBALLE, T.H., and CHILTON, F., Effects of Plastic Deformation on the Superconducting Specific-Heat Transition of Niobium, *J. Appl. Phys.* 50, 6423-36 (1979).

[ZWI70] ZWICKER, U., LÖHBERG, R., and HELLER, W., Metallkundliche Probleme und Supraleitung bei Legierungen auf Basis Titan-Niob, die als Werkstoffe für die Herstellung von supraleitenden Magneten dienen können, *Z. Metallkde.* 61, 836-47 (1970).

[ZWI74] ZWICKER, U., *Titan und Titanlegierungen,* Springer-Verlag, 1974.

Symbols and Abbreviations

A elastic anisotropy ratio; $A = C_{44}/C'$

Å   angstrom unit; $1\,\text{Å} = 10^{-10}$ m

AC   alternating current, alternating

ARR   area-reduction ratio, A_i/A_f, in rod or wire drawing

a lattice parameter

at.%   atomic percent; "A-B(n at.%)" indicates n at.% of element B dissolved in A. The composition of a multicomponent alloy in atomic percent is represented by A_l-B_m-C_n, etc.

BCS   Bardeen, Cooper, and Schrieffer [Bar57] theory of superconductivity

bcc   body-centered cubic

bracket notation   ( ) lattice plane
{ } generalized lattice plane
[ ] crystallographic direction
⟨ ⟩ generalized crystallographic direction

C specific heat

C_e electronic component of the specific heat of a metal; $C_e = \gamma T$

C_{es} electronic specific heat of superconducting electrons

C_{ij} elastic stiffness modulus (elastic constant) in Voigt notation

C' elastic shear stiffness modulus; shear constant; $C' = (C_{11} - C_{12})/2$

CPA   coherent potential approximation

c concentration in general; solute concentration

c_A concentration of component A in a mixture

c-axis   the hexad axis in a hexagonal-close-packed (hcp) crystal

D solute or tracer diffusion coefficient (diffusivity)

D_i diffusivity of element i

D_0 frequency factor, or prefactor, in the diffusivity expression

d diagonal of a hardness imprint in H_V measurement

d diameter of a deformation cell (subband)

d spectroscopic state of valence electrons

E Young's modulus

E energy of a band electron

E_F energy of a band electron at the Fermi level; Fermi energy

ELI   extra-low-interstitial (grade of Ti)

EDAX   energy dispersive analysis of x-rays

eV   electron-volt

e/a electron/atom ratio

f resonant frequency of a vibrating rod

f volume fraction, mole fraction, etc.; frequently with subscripts; $0 < f < 1$

f_R	fractional change in electrical resistance; used when following the progress of a metallurgical aging reaction resistometrically	M_d	temperature at the start of deformation-martensitic transformation
fcc	face-centered cubic	M_f	temperature at the finish of spontaneous martensitic transformation
G	elastic shear modulus		
$G_{H,R,V}$	Hill-, Reuss-, and Voigt-calculated, respectively, elastic shear moduli	M_s	temperature at the start of spontaneous martensitic transformation
GN	group number; electron/atom ratio of an unalloyed transition metal	m	meter(s)
		min	minute(s)
g	free energy per unit mass	$\mathfrak{N}$	Avogadro's number, 6.025×10^{23} molecules/mole
H	magnetic field strength		
H_a	strength of an applied magnetic field	N_{eff}	an "effective" electron/atom ratio
H_V	Vickers hardness	NFE	nearly free electron (electronic-structure model)
H_{c1}	lower critical field of a type-II superconductor	$n(E)$	density of electron states of energy E (number per unit energy)
H_{c2}	upper critical field of a type-II superconductor		
h	hour(s)	$n(E_F)$	Fermi density of states; the above at $E = E_F$
h	Planck's constant	O_{eq}	oxygen-equivalent concentration; concentration of oxygen having the same strengthening effect as all the interstitial elements actually present
hcp	hexagonal-close packed		
K	bulk modulus		
$K_{H,R,V}$	Hill-, Reuss-, and Voigt-calculated, respectively, bulk moduli		
KKR	electronic-structure calculational procedure of Korringa, Kohn, and Rostoker	OPW	orthogonalized plane wave
		p	spectroscopic state of valence electrons
k_B	Boltzmann's constant	Q	a texturization parameter based on relative paramagnetic anisotropy in magnetic texture measurement
ksi	10^3 pounds per square inch; a common unit for tensile stress and yield stress		
L	load imposed on a hardness indenter in H_V measurement	r	cooling rate during quenching
		r_c	critical cooling rate, faster than which M_s is r-independent
l	length of a rod or other object		
l	liquid phase in Ti-alloy phase diagrams	S_{ij}	elastic compliance modulus in Voigt notation
M	magnetization, i.e., magnetic moment per unit volume; $4\pi M$ is also the magnetization	SM	"simple metal", such as Al, Sn, etc.; may also represent a metalloid such as Ge, Si, etc.
M	mean atomic weight of an alloy	SAD	selected-area diffraction; a special technique of transmission electron microscopy

SAXS	small-angle x-ray scattering (diffraction)		The composition of a multi-component alloy in weight percent is represented by $A\text{-}mB\text{-}nC$, etc.
STEM	scanning transmission electron microscopy		
s	spectroscopic state of valence electrons	Y	yield strength (usually $Y = \sigma_{0.2}$)
T	temperature	Y_{ULT}	ultimate strength; maximum
T_c	superconducting transition temperature		stress in the stress-strain curve. In materials with limited ductility, Y_{ULT} also
T_0	some thermodynamic-equilibrium temperature		equals σ_B, the stress at fracture
T_{OD}	order-disorder transition temperature		
TB	tight-binding (electronic structure calculational procedure)		
TM	transition metal		**Greek Symbols**
TEM	transmission electron microscopy	α	hexagonal-close-packed (hcp) crystal structure in Ti-base alloys
T-T-T	"time, temperature, and transformation"; a type of diagram representing the kinetics of phase transformation	α'	solute-lean, hcp, martensite
		α''	solute-rich, orthorhombic, martensite
t	some "reduced" (or normalized) temperature	α^m	generic martensite (includes α' and α'' indiscriminately)
t	time	α_2	an ordered hexagonal phase
V	atomic volume; volume of unit cell divided by the number of atoms in it; used with subscripts β or ω		(DO_{19} in Ti_3Al, Ti_3Ga, Ti_3In, and Ti_3Sn) of composition near Ti_3SM in Ti-SM alloys
VRH	Voigt, Reuss, and Hill elastic-constant-based macroscopic modulus (E, G, and K) calculational method	β	lattice specific-heat coefficient
		β	body-centered-cubic (bcc) crystal structure in Ti-base alloys
VRHG	Voigt, Reuss, Hill, and Gilvarry elastic-constant-based Debye-temperature calculational method	β'	a solute-lean β-phase
		β''	a solute-rich β-phase
		β_s	start of the $\alpha^m \to \beta$ reversion on up-quenching
v_{sub}	vibrational- or acoustic-wave velocity; sub = L (longitudinal), T (transverse), *torsion* (torsional), m (mean)	β_{tr}	the product of a transformation from the β phase. If the transformation product is Widmanstätten, as is usually the case, it is referred to as $(\alpha + \beta)_W$
WQ	quenched into water		
wt.%	weight percent; "$A\text{-}B(n$ wt.%)" indicates n wt.% of element B dissolved in A; $A\text{-}nB$ indicates the same.	γ	an intermetallic compound (generic) in Ti-alloy binary phase diagrams

γ an ordered face-centered-tetragonal compound ($L1_0$) of composition near TiAl in Ti-Al alloys

γ electronic specific-heat coefficient; $C_e = \gamma T$

γ' Ni_3Al precipitate in Ni-base superalloys

δ an electron-scattering-strength parameter in alloy theory; $0 < \delta < 1$

δ_B elongation of a tensile-test specimen just prior to fracture

ϵ strain, percent deformation

θ an angle of rotation

θ_D Debye temperature

λ one of the two Lamé elastic-modulus parameters

μ one of the two Lamé elastic-modulus parameters

μm   micrometer, micron, 10^{-6} m

ν Poisson's ratio

ρ electrical resistivity

ρ_d (mass) density

ρ_i "ideal" component of electrical resistivity due to electron-phonon scattering

ρ_s "impurity" component of electrical resistivity due to electron-impurity (solute) scattering

σ stress

σ_B fracture strength. In materials with limited ductility, $\sigma_B = Y_{ULT}$

σ_f flow stress

$\sigma_{0.01}$ the stress at 0.01% strain; the proportional limit

$\sigma_{0.2}$ the stress at 0.2% strain; the "0.2%-offset" yield stress

ϕ as superscript, this signifies the diameter of a wire, small sphere, electron beam, etc.

ϕ_c a texturization parameter in magnetic texture measurement

χ magnetic susceptibility

χ_A magnetic susceptibility of component A in a mixture

$\chi_\parallel$ component of total magnetic susceptibility of an hcp crystal directed along the hexad axis

$\chi_\perp$ component of total magnetic susceptibility of an hcp crystal lying within the basal plane

χ_{sub} magnetic susceptibility component according to: sub = P (Pauli spin paramagnetism), L (Landau diamagnetism), so (spin-orbit component), orb (orbital paramagnetism), i (ion-core diamagnetism)

Ω_{ij} thermodynamic interaction parameter between elements "i" and "j"

ω ω phase

Titanium-Alloy Index

(see also Table of Contents)

Within the various groupings (binary, ternary, etc.), alloys are listed in alphabetical order. In the case of multicomponent alloys, the order of the ingredient-string has been previously dictated nonalphabetically by considerations such as: order of importance of the ingredients, α- or β-stabilizing tendencies of the ingredients, commercial practice, etc. In the case of commercial alloys and some others, compositions are indicated; otherwise, the listings refer indiscriminately to single alloys or a series of them. Page numbers in parentheses ( ) indicate figures; those in square brackets [ ] indicate tables.

239

Subject Index

Entries beginning with the Greek symbols α, β, γ, and ω precede the A, B, C, and Z entries, respectively.

242